UNIVERSAL LAWS *of* NATURE AND CELLS

A NEW APPROACH

GORAN INDJIC

UNIVERSAL LAWS OF NATURE AND CELLS
A NEW APPROACH

iUniverse books may be ordered through booksellers or by contacting:

iUniverse
1663 Liberty Drive
Bloomington, IN 47403
www.iuniverse.com
1-800-Authors (1-800-288-4677)

Because of the dynamic nature of the Internet, any web addresses or links contained in this book may have changed since publication and may no longer be valid. The views expressed in this work are solely those of the author and do not necessarily reflect the views of the publisher, and the publisher hereby disclaims any responsibility for them.

Any people depicted in stock imagery provided by Getty Images are models, and such images are being used for illustrative purposes only. Certain stock imagery © Getty Images.

ISBN: 978-1-5320-6837-9 (sc)
ISBN: 978-1-5320-6836-2 (e)

Library of Congress Control Number: 2019903262

Print information available on the last page.

iUniverse rev. date: 03/27/2019

About the Universal Laws of Nature and Cells

The goal of this book is to offer new ideas and experiments for investigating prokaryotic and eukaryotic cells and new treatments for infectious and malignant diseases by technical devices and other means.

The complex and chaotic opinions in the literature about structures and different phenomena of prokaryotic and eukaryotic cells, the poor results of the existing treatment for infectious and malignant diseases, the huge number of unexplored prokaryotic cells, and the complex taxonomy of prokaryotic cells initiated the idea of searching for simple explanations for the different structures and phenomena of prokaryotic and eukaryotic cells by the universal laws of nature.

The observation of the structures and the different phenomena of prokaryotic and eukaryotic cells in causative relationships according to the universal laws of nature produced the new approach to structures and different phenomena of prokaryotic and eukaryotic cells.

Prokaryotic and eukaryotic cells consist of polypeptides that attract coenzymes and nonprotein hormones that are present only in eukaryotic cells according to the new approach. Coenzymes and nonprotein hormones attract metabolites that bring electrical charges. Polypeptides of prokaryotic and eukaryotic cells attract each other along their lengths by opposite electrical charges of metabolites and other ions and build protein spirals.

Protein spirals attract each other in helicoid pairs that have the same structure as ds DNA building the strings that attract ds DNA and the other nucleic acids in them. Protein spirals move in opposite directions in the strings by each other performing synthesis and decomposition of metabolites including biosynthesis of polypeptides, nucleotides, and nucleic acids.

Strings build filaments of prokaryotic cells and complex cylinders of eukaryotic cells that enable their division. Filaments build different structures of prokaryotic cells (flagella, pili, nucleoid, membrane, and others). The complex cylinders build different structures of eukaryotic cells (nucleus, chromosomes, the Golgi apparatus, endoplasmic reticulum, mitochondria, and others).

Filaments and complex cylinders are new structures found and described for the first time in the literature by described structures and phenomena of prokaryotic and eukaryotic cells. The proof of their existences is that different phenomena and structures of prokaryotic and eukaryotic cells cannot exist in causative relationships without them; they rely on universal laws of nature.

Different characteristics of prokaryotic and eukaryotic cells including malignant ones depend on different lengths of polypeptides, lengths and numbers of filaments, and complex cylinders. The different lengths of polypeptides, lengths and numbers of filaments, and complex cylinders attract quantities of coenzymes and nonprotein hormones for eukaryotic cells. The evolution of prokaryotic and eukaryotic cells and their predecessors is a product of a continuous change of environment.

The new approach to different phenomena of prokaryotic and eukaryotic cells produced a new taxonomy of prokaryotic cells, a universal method for investigating different prokaryotic cells, a new understanding of differentiation of stem cells, and a new understanding of pathogenesis of malignant cells. It also revitalized the old idea of treating infectious and malignant diseases by technical devices.

The new approach to prokaryotic and eukaryotic cells according to the laws of nature contradicts the complex approach of contemporary literature that deals with prokaryotic and eukaryotic cells but not with the phenomena that exists according to universal laws. The new approach to prokaryotic and eukaryotic cells according to universal laws of nature is also imperfect, and it looks for criticism to progress.

This book will succeed only if readers criticize the text and produce a new approach to different phenomena of prokaryotic and eukaryotic cells and new ideas for treatment of infectious and malignant diseases— that is, if readers negate this book according to universal laws in a new approach.

About the Author

Goran Indjic had an opportunity as a physician and a clinical microbiologist to examine patients with infectious and malignant diseases working in hospitals and health care systems in Europe and Africa.

He identified pathogens in university microbiology labs with technologists, and he suggested antimicrobial treatments and preventive measures for hospital-acquired infections. He taught medical students theoretical and practical medical microbiology in South Africa. In Canada, he worked in a veterinary lab as a microbiologist identifying different pathogens in animals. He performed experiments on prokaryotic cells that proved their different characteristics were subject to laws of nature. The result is a new approach to phenomena of prokaryotic and eukaryotic cells according to universal laws of nature presented in the book.

He also designed a new health care system by electronically unifying medical records in poor countries.

His focus is on explaining the structure of prokaryotic and eukaryotic cells and other phenomena according to the universal laws and life in general. He believes human beings are divine creatures who must fight stupidity, confusion, and poverty by all means. He believes also that art and creativity in any field are bread and water for all human beings.

Preface

Human beings exist by understanding the cruel laws of nature and using them in devices for surviving or killing one another. Good examples of the cruelty of nature are malignant diseases caused by cancer cells and infectious diseases caused by prokaryotic cells and viruses.

Malignant and infectious diseases do not have a clear pathogenesis and effective treatments because we do not understand how the laws of nature produce the characteristics of different cells and do not have the technical means to suppress their growth.

The universal laws of nature also produce characteristics of prokaryotic and eukaryotic cells, including cancerous ones. If the universal laws of nature are connected with basic structures of prokaryotic and eukaryotic cells and their metabolism, stories about them are not chaotic and boring as contemporary literature describes them.

Different characteristics of prokaryotic and eukaryotic cells have not been explained according to universal laws because their basic structures are not crystallized and connected with their metabolism and explained by each other despite rich literature about them. Basic structures were crystallized, explained, and connected with metabolism of prokaryotic and eukaryotic cells before explanations of their characteristics by universal laws came to be.

According to the new approach, basic structures of prokaryotic and eukaryotic cells consist of polypeptides that build protein and nucleic acid spirals, which in turn build strings that build filaments of prokaryotic cells and complex cylinders of eukaryotic cells.

The strings, filaments, and complex cylinders are structures of the cells I describe for the first time in the literature. Filaments and complex cylinders build and unify their structures by metabolism. Structures of prokaryotic and eukaryotic cells are not described in the literature based on their real nature because they are more like structures of coagulated proteins. This was because prokaryotic and eukaryotic cells were observed in dead cells without any deeper analysis, imagination, or understanding.

Their picture changes steadily despite this because of logical approaches many scientists take. My descriptions of prokaryotic and eukaryotic cells are logical, simple, and subjective; I built them on facts and knowledge about them, imagination, and logical thinking.

A subjective approach to prokaryotic and eukaryotic cells is not good if it contradicts common sense and their real nature. Any subjective and logical picture of them is better than any objective but chaotic one without common sense because new facts can easily destroy chaotic one. Any objective and chaotic pictures of them in contemporary biology usually cause confusion because they do not have the logic of cause and effect in their phenomena, so new facts do not disturb them.

I analyzed many phenomena of prokaryotic and eukaryotic cells described by biologists. Any pictures of them are chaotic information that needs analysis and explanation.

More research on this new approach to prokaryotic and eukaryotic cells is needed to find their real nature, but knowledge of them is never complete because of our limited nature.

This book will open new frontiers concerning prokaryotic and eukaryotic cells and will achieve its goal if readers replace it with something more logical and practical. I intended to inspire scientists from different disciplines to make new discoveries in microbiology that will combat infectious and malignant diseases.

I dedicate this book to everyone who is puzzled by living forms and wants to develop new treatments for diseases or discover more about prokaryotic and eukaryotic cells.

I express my gratitude to my family and the many people who directly or indirectly were involved in this work.

Goran Indjic
Toronto, 2017

Contents

Introduction

The book is about new structures of prokaryotic and eukaryotic cells, different phenomena of prokaryotic and eukaryotic cells, and universal laws of nature. The phenomena of prokaryotic and eukaryotic cells and universal laws designed, explained, and proved the new structures of prokaryotic cells.

New structures connected and explained the different phenomena of prokaryotic and eukaryotic cells in causative relationships. These structures and phenomena are inseparable as is the case with other phenomena; the structures consist of polypeptides and different coenzymes that attract different metabolites as ions. The different polypeptides, coenzymes, metabolites, and nucleotides build protein and nucleic acid spirals that attract each other and build strings that build filaments of prokaryotic cells and complex cylinders of eukaryotic cells.

The basic polypeptides, strings, filaments, and complex cylinders are new structures designed to explain organelles, metabolism, binary fission, mitosis, meiosis, and other phenomena of prokaryotic and eukaryotic cells according to the laws of nature.

The book criticizes scholastic approaches to prokaryotic and eukaryotic cells because their conclusions produced much dogma that hindered greater understanding of prokaryotic and eukaryotic cells. Scientists accept information about prokaryotic and eukaryotic cells without deeper analysis, understanding, and imagination according to the universal laws of nature and wrap their thoughts in empty words and phrases because they did not connect them in causative relationships (appendix I). Different phenomena of prokaryotic and eukaryotic cells exist in nature only in causative relationships.

The central dogma of protein biosynthesis does not explain the growth of prokaryotic and eukaryotic cells in a natural way and the roles different structures play in it. This new approach to protein and nucleic acid biosynthesis replaces the central dogma of protein biosynthesis and can explain the origin and the growth of different structures and metabolism of prokaryotic and eukaryotic cells.

According to new research, protein and nucleic acid biosynthesis is part of a unified system of phenomena that depend on each other in a cause and effect relationship. Many biologists lost interest in learning more about the real nature of prokaryotic and eukaryotic cells for many reasons including chaotic experiments on them that produced only more chaos, petrified dogma, and absolute truth without regard for the real nature of the cells. They stopped understanding their nature and coming up with useful inventions based on that (appendix II).

Many biologists did not understand the clear connection between polypeptides and protein structures and their movements, protein and nucleic acid biosynthesis, metabolism, and other phenomena of prokaryotic and eukaryotic cells, and this is critical if the biological sciences want to develop new treatments for infectious and malignant diseases.

The structures and the phenomena of these cells are a unified system of cause and effect based on the universal laws of nature, which explain the characteristics of the cells based on the quantities of coenzymes of the protein structures and their evolution (see appendix III).

Great people in history used the universal laws of nature to explain many secrets of nature and to produce divine works of science and art. Heraclitus (535–475 BC) noticed them in nature: "Everything flows and nothing stays." Michelangelo (1475–1564) used them intuitively as did J. S. Bach (1686–1750) to harmonize God, the stars, and life.

Antoine Laurent Lavoisier (1743–1794) created and proved the law of conservation of matter in chemical reactions; he proved that matter was indestructible though it changed natures. Charles Darwin (1809–1882) explained and confirmed evolution of different forms of life by them though he did not mention them (appendix IV).

Marx (1818–1883) and Engels (1820–1895) defined them in philosophy as the law of dialectical materialism (appendix III). Dmitri Ivanovich Mendeleev (1834–1907) formulated the periodic law and with it confirmed their existence. Nikola Tesla (1856–1943) explained many electrical phenomena and produced many inventions based on his understanding of the essential characteristics of matter: "If you want to find the secrets of the universe, think in terms of energy, frequency, and vibration."

Albert Einstein (1879–1955) introduced them to physics with his famous formulation $E = mc^2$. He thought light traveled the fastest while Tesla thought that the ether through which he thought light traveled was faster.

Matter and energy cannot exist separately because they are essentially the same. Space cannot exist without matter, which takes on different qualities and quantities over time but remains. They come from different forms of matter and will become different forms of matter according to the laws of dialectic materialism. Heraclitus wrote, "No man ever steps in the same river twice, for it is not the same river and he is not the same man."

The universe has always existed and will always exist because time is inseparable from matter. Matter's future forms rely on its past forms. Watson (1928–) and Crick (1916–2004) determined that every organism was created by something in a parent organism.[33] They wrote, "It has not escaped our notice that the specific pairing we have postulated immediately suggests possibly a coping mechanism for the genetic material."

Watson and Crick revolutionized biology by introducing the structure of ds DNA in it though they did not understand how the universal laws of nature produced different organisms. The facts are that ds DNA as complex helicoid molecules develop different forms of life; it produces polypeptides among polyribosomes during polypeptide biosynthesis described by the central dogma, which is accepted as truth without logical explanations or alternative approaches. Rosalind Franklin (1920–1958) produced the facts about ds DNA that were necessary for Watson's and Crick's conclusions.[33]

No doubt the laws of dialectical materialism or the universal laws of nature enlighten the nature of prokaryotic and eukaryotic cells though many people connect the laws of dialectical materialism with politics and stupidities that happened in communist countries, but "Stupidity is cosmic power" (Miroslav Krleza, 1893–1988).[34]

I wrote this book for a practical reason: several hundred thousand if not millions of prokaryotic cells have yet to be investigated. The new approach to prokaryotic cells and the laws of dialectic materialism enabled designing a new taxonomy of prokaryotic cells and universal methods for their identification. This will enable investigation of bacterial ecosystems and their species and creation of artificial ones for industrial, agricultural, and medicinal needs.

This new approach to eukaryotic cells and the laws of dialectic materialism enabled a new understanding of cells, stem cell differentiation, mitosis, meiosis, and the pathogenesis of malignant cells. Experiments on eukaryotic cells will enable regulation of cancer cells and killing cells infected with viruses and bacteria.

This book's appendices offer more information on prokaryotic and eukaryotic cells and the research that has been done on them as well as a new method and proposals for new experiments.

CHAPTER 1
STRUCTURE OF PROKARYOTIC CELLS

Past and present researchers have dealt with prokaryotic cells, but in many cases, their investigations were irrational and inconclusive.

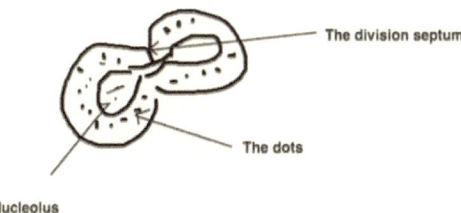

Electron micrographs of thin sections of *Neisseria gonorrhoeae* do not show the cytoskeleton inside but compact structures with dots.

According to a new approach, prokaryotic cells consist of filaments of different lengths and diameters and different qualities of protein. Filaments attract each other and form compact structures—cells, flagella, plasma membranes, plasmids, and nucleoids—identifiable in electron micrographs.

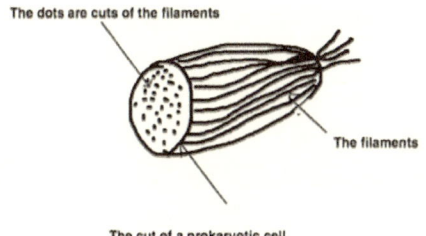

Dots in thin sections of the electronic images of prokaryotic cells are cuts of filaments.

Prokaryotic cells are still dogmatically considered in the literature as bags of proteins, minerals, and other molecules floating in water due to the misleading images of the electronic microscope, misleading facts, and a lack of imagination on the part of researchers (appendix V). Luckily, the concept of the inner structure of prokaryotic cells is changing due to investigation of their cytoskeleton and other facts. Literature started to accept that the cytoskeleton consists of filaments despite many strange models of flagella (appendix VI). According to the new approach, different filaments build prokaryotic cells as unified systems of cell membranes, nucleoid, plasmids, pili, and flagella (appendix VII). Filaments can be seen as intertwined structures covered by cell walls of some prokaryotic cells in electronic images.

The parts of the filaments

The electronic images of *Campylobacter jejuni* show intertwined filaments covered with the cell wall.

A flagellum (a pilus) is part of a filament according to the new view explained here. Filaments grow at opposite ends and must have synchronized divisions that cause binary fissions of prokaryotic cells. Each filament has two strings with DNA that are open on the ends. The filaments' strings interchange parts of protein and nucleic spirals. The circular chromosome does not exist in prokaryotic cells according to the new approach. The nucleic acids are disconnected and spread in filaments along strings also according to the new approach. The analysis and synthesis of the facts and views of the contemporary literature and their contradictions are put in the appendices. They explain a logic of this new approach.

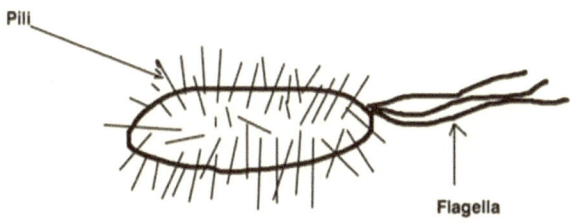

Flagella and pili are parts of the filaments with different thicknesses and lengths.

Filaments

According to the new approach, filaments must have specific organization that enables them to grow at opposite ends (see picture below). Each string in a filament must have a straight and a coiled part. Filaments have two straight parts with two coiled ones around them on opposite sides. The straight and coiled parts attract each other by metabolites. The coiled part of a string grows whereas the straight part does not. This organization of filaments enables them to grow in opposite directions; their division and binary fission of prokaryotic cells is a consequence.

The model of filament has the two strings that have specific structures—coiled and straight parts. This new model is called the Rosalind Franklin model because the protein spirals of the strings have the organization of ds DNA (appendix VIII).

A filament string must consist of the two protein spirals that have a ds DNA structure.

The protein spirals consist from the polypeptide spirals

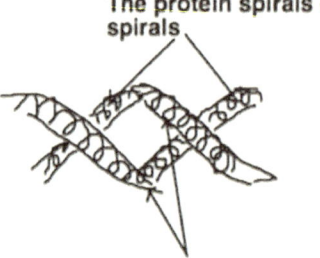

The protein spirals have the rope or helicoid structure

This picture shows a string with two protein spirals that contain polypeptide spirals; they consist of polypeptides that attract each other along their lengths.

The two protein spirals must have the rope structure of DNA to attract strands of double-stranded DNA. The other nucleic acids that appear during biosynthesis of polypeptides and nucleic acids must have specific positions in the polypeptides of protein spirals of strings so biosynthesis of polypeptides and nucleic acid spirals can take place according to nature's logic. The positions of nucleic acids are explained in protein and acid spirals synthesis.

The protein spirals have the same helicoid structure as ds DNA

The model of the string has protein spirals and ds DNA.

The protein spirals must have polypeptide spirals that attract nucleic acid spirals of ds DNA and metabolism.

The protein spiral

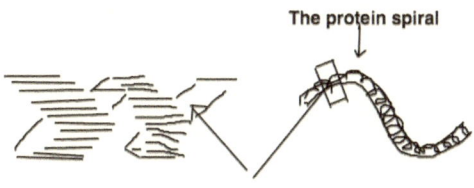

The polypeptide spiral

Protein spirals consist of polypeptide spirals made up of polypeptides, which have different qualities and lengths by which they attract different concentrations of coenzymes and metabolites, which create the metabolic characteristics of different prokaryotic species.

Polypeptides must have a spiral structure due to circular movement of amino acids around their bonds. Polypeptide spirals attract each other by metabolites along their lengths and move each other by spiral movement that perform metabolism. Filaments without these structures cannot metabolize different molecules simultaneously. I explain how prokaryotic cells metabolize different molecules in the next chapters.

Polypeptides and Coenzymes

The specific polypeptides of different lengths in the spirals attract specific coenzymes (NAD, FMN, ATP, B_1, folic Acid, biotin, CoA, vitamin K, and B_6). Specific coenzymes attract specific metabolites based on their electrical charges to form polypeptide spirals. Amino acids of polypeptides of polypeptide spirals attract each other by hydrogen bonding.

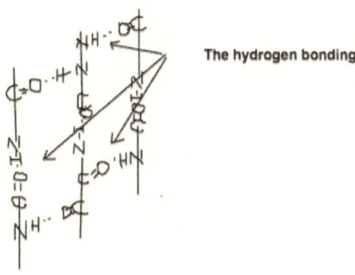

The hydrogen bonding

Amino acids of polypeptides of polypeptide spirals attract each other along the lengths and tops by hydrogen bonding, which makes polypeptide spirals more stable.

The same metabolites appear during decomposition and synthesis of simple sugars, fatty acids, amino acids, nucleotides, coenzymes, and other molecules. The decomposition and synthesis of the different molecules during metabolism of prokaryotic cells are explained later with more details.

The two wheels depict the cuts of the two close polypeptide spirals that attract each other during metabolism. The teeth of the wheels depict the cuts of the specific polypeptides with coenzymes that attract specific metabolites.

Polypeptides in polypeptide spirals have specific organization of coenzymes that attract metabolites as ions during metabolism. Polypeptides attract each other by opposite electrical charges of their ions attracted by coenzymes. This attraction enables three very important phenomena of filaments simultaneously: movement of spirals in opposite directions, decomposition and synthesis of metabolites, and formation of polypeptides of protein spirals and pieces of nucleic acid spirals of strings of filaments and growth.

The growth of filaments causes binary fissions of prokaryotic cells. All phenomena are explained later after explanations of polypeptides with coenzymes that produce peptidoglycan, a universal molecule of many prokaryotic cells. This molecule's structure is very important because it can discover the structure of coenzymes and their metabolites attracted by polypeptides. This molecule to be made looks for structures of coenzymes attracted by their polypeptides.

Specific Polypeptides

Structure of specific polypeptides with specific coenzymes of the proteins spirals can be built and proven by the structure of NAG and NAM acid. Polypeptides synthesize them by adding metabolites to glucose molecules (appendix IX). NAG and NAM acids are molecules that appear alternatively in chains of peptidoglycan complex molecules.

The next conclusion ensued from the chains with NAG and NAM acid molecules: the specific coenzymes groups of the specific polypeptides that produce them must also repeat alternatively along the polypeptide spirals to produce them.

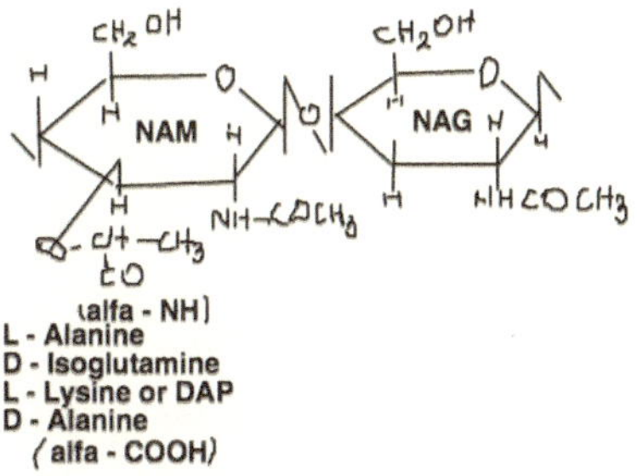

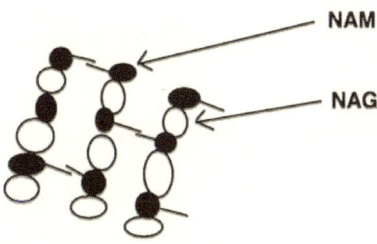

(alfa - NH)
L - Alanine
D - Isoglutamine
L - Lysine or DAP
D - Alanine
(alfa - COOH)

NAM has short polypeptides. NAM and NAG molecules repeat alternatively in their chains.

Different prokaryotic cells have NAM acids with short polypeptides that connect chains with alternative NAG and NAM acid molecules into peptidoglycan nets.

NAM

NAG

Chains with NAG and NAM acid are connected in the net of the huge peptidoglycan molecules at their NAM acids by specific polypeptides.

Some prokaryotic cells in their short polypeptides have diaminopimelic acid (DAP) instead of L-lysine. The short polypeptides of NAM acids with L-lysine connect with each other by peptides that have five glycine molecules into peptidoglycan. The short polypeptides of NAM acids with diaminopimelic acid connect each other by their diaminopimelic acids also into peptidoglycan.[1]

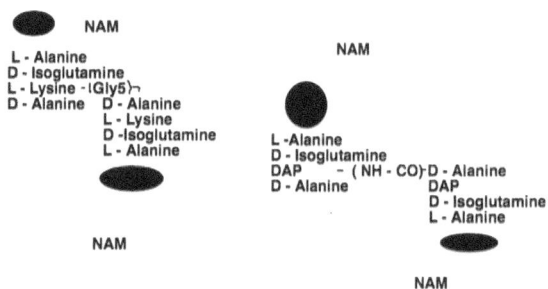

This shows different connections between the short polypeptides of NAM acids into peptidoglycan.

Prokaryotic cells have different cells walls. Some prokaryotic cells have cell membranes and thick peptidoglycan. These prokaryotic cells are known as gram-positive bacteria. Some prokaryotic cells have a cell membrane, a thin peptidoglycan, and an outer membrane. These prokaryotic cells are known as gram-negative bacteria.

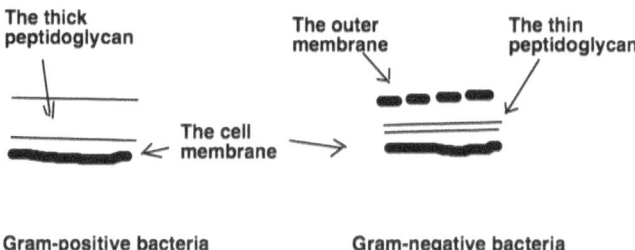

The rough explanation of the structures of prokaryotic cells looks for a rationale before explanation of the model of specific polypeptides and new structures connected with movements and metabolism. The rationale is a set of reasons or a logical basis for a course of action or particular belief according to the dictionary.

Chapter 2
Rationale for New Structures of Prokaryotic Cells

Polypeptides with coenzymes, polypeptide spirals, protein spirals, strings, and filaments were constructed by contemporary views of the structures and metabolism of prokaryotic cells and the facts and many possibilities of their connections during metabolism and growth.

The understanding of the new structures of prokaryotic cells was a slow process because the new structures had to satisfy metabolism of the molecules and other phenomena of prokaryotic cells. The phenomena in causal relationships prove the new structures of prokaryotic cells because specific phenomena cannot exist without new structures. Process of the understanding of the structures and the phenomena of prokaryotic cells is summarized below.

- The selectively permeable nature of prokaryotic membranes keeps ions, proteins, and other molecules in the cell. Small molecules and ions move through the membrane by osmosis while big molecules move through membrane by the active transport according to contemporary literature. This approach is dogmatic because it does not explain the deeper connection of prokaryotic membranes with other structures and the metabolism of the different molecules (see appendix X).
- Prokaryotic cells have different concentrations of proteins with different concentrations of the same coenzymes, the same metabolites, and the same complex molecules of metabolism. This conclusion ensued from the fact that different prokaryotic cells have different quantities of the same and different molecules.

- Different proteins of prokaryotic cells are also different enzymes that attract the same coenzymes. Different enzymes attract different metabolites as substrates that bring the same specific group of atoms attracted by the same specific coenzymes. The coenzymes of different enzymes attract metabolites (the same group of atoms) from metabolites during metabolism. This conclusion ensued from the fact that prokaryotic cells have different enzymes with a similar organization, the same coenzymes, and the same metabolites as ions.

- Enzymes consist of specific proteins and coenzymes that attract metabolites as ions. Enzymes catalyze biochemical reactions of attracted molecules making from them smaller metabolites during decomposition or bigger ones during synthesis.[2] Enzymes are classified according to international classifications (see appendix XI).

- Prokaryotic cells decompose and synthesize metabolites simultaneously with precedence of decomposition or synthesis depending on their source of energy. These conclusions ensued from the fact that photosynthetic bacteria produce more-complex organic compounds than decompose them because they absorb the sun's energy and store it in ATP and complex organic molecules (fats, polysaccharides, proteins, and other molecules). Nonphotosynthetic bacteria that use biochemical energy from complex molecules decompose complex organic molecules rather than synthesize them to gain energy for metabolism.

- NAG and NAM acids in peptidoglycan molecules are common molecules among many prokaryotic cells, so bacteria with peptidoglycan and different enzymes must have the same structures of the coenzymes in different enzymes to produce the same structure of peptidoglycan molecules.

- Groups of coenzymes attracted by polypeptides of different proteins build NAG and NAM acids by their metabolites. Polypeptides in coenzymes add specific molecular groups to glucose molecules in polysaccharide chains transforming them into NAG and NAM acid of peptidoglycan. This conclusion ensued from the fact that only specific enzymes with specific coenzymes can build them.

- Polypeptides and coenzymes attract metabolites with electrical charges during metabolism and produce bioelectricity. Bioelectricity of prokaryotic cells is a consequence of the movement of metabolites among coenzymes of different polypeptides of protein spirals during metabolism. These conclusions ensued from the facts that electricity is the movement of electric particles through a conductor and that bioelectricity is the movement of metabolites with electrical charges among coenzymes through proteins during metabolism.

- Proteins in prokaryotic cells must move to produce metabolism, bioelectricity, and growth. These conclusions ensued from the fact that metabolites, coenzymes, and protein structures produce bioelectricity and from the facts of the growth of different structures of prokaryotic cells.

- The quantities of polypeptides in proteins of prokaryotic cells attract gmol/l concentrations of the same coenzymes and the same metabolites. The different concentrations of the same metabolites must be attracted by different concentrations of coenzymes attracted by different enzymes. Otherwise, different characteristics of prokaryotic cells could not exist. The conclusion ensued from the fact that different prokaryotic cells have different concentrations of the same metabolites and different proteins.

- Coenzymes attract specific metabolites from each other among different enzymes (proteins) during metabolism according to the strict connections. The conclusion ensued from the fact that different prokaryotic cells have the same pathways for metabolites attracted by coenzymes.

- Prokaryotic cells have filaments built from proteins of unknown organization. The conclusion ensued from the fact that prokaryotic cells have cytoskeletons that consist of filaments with very similar metabolism and different proteins. This approach is also accepted by contemporary literature (see appendix XII).

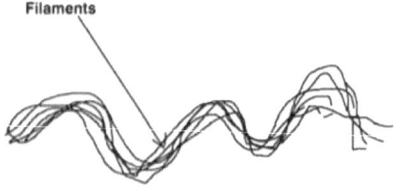

Electron microscope picture shows filaments in spiral prokaryotic cells.[3]

More Conclusions

- Prokaryotic cells have different concentrations of different proteins that attract different concentrations of coenzymes and metabolites but different complex molecules. Different concentrations of different enzymes with different concentrations of the same coenzymes attract different concentrations of the same metabolites.
- Concentrations of different proteins and enzymes, the same coenzymes and metabolites and different complex molecules produce different characteristics of prokaryotic cells and similar phenomena. The conclusion ensued from the facts that prokaryotic cells have different concentrations of proteins, enzymes, coenzymes, the same metabolites and different complex molecules and that prokaryotic cells have the similar phenomena (metabolism, growth, binary fission, and others).
- Prokaryotic cells must have the same organization of specific groups of molecules attracted by coenzymes in polypeptides of different spiral proteins of filaments as the groups of atoms that belong to NAG and NAM acid to synthesize them. The same organization of specific groups of molecules attracted by groups of coenzymes in polypeptides of protein spirals of filaments performs movement of protein spirals and filaments together with the metabolism of metabolites.

- Protein structures, metabolism, and phenomena of prokaryotic cells can exist together in only cause-and-effect connections; they exist as separate entities only in contemporary literature.
- If specific coenzymes exist only in long, linear polypeptides without polypeptide spirals, long and linear polypeptides must have the same concentration of coenzymes and metabolites that repeat to make strict structures of NAG and NAM acid molecules of peptidoglycan. This hypothesis is rejected because different prokaryotic cells have different concentrations of basic metabolites attracted by different concentrations of coenzymes.
- If coenzymes exist in short polypeptides (as opposed to long and linear polypeptides) and if polypeptides attract each other along their lengths by their metabolites (ions) and build polypeptide spirals in prokaryotic cells, polypeptides can have specific and different concentrations of coenzymes and their metabolites to build NAG and NAM acids and have different metabolisms with different concentrations of metabolites simultaneously.

Only polypeptides attracted along their lengths in polypeptide spirals can solve the contradiction of specific and different concentrations of coenzymes and the same metabolites at different prokaryotic cells. This logical hypothesis is taken as the model for the building of polypeptide and protein spirals that build filaments. The existence of the new structures in prokaryotic cells is proven by the existence of different metabolisms at different prokaryotic cells that produce the same metabolites of different concentrations. The existence of the new structures inside prokaryotic cells is also proven by the existence of peptidoglycan molecules that look for specific concentrations of the same metabolites for their production.

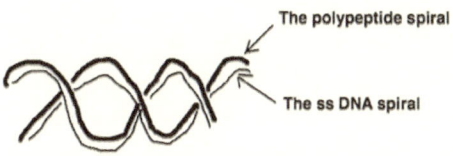

The polypeptide spiral

The ss DNA spiral

The polypeptide and ds DNA spirals must have the helicoid structures to attract and hold each other.

- Polypeptide spirals cannot be linear to attract and hold ds DNA with two helicoid strands and provide synthesis and decomposition of different metabolites simultaneously. Polypeptide spirals must exist as the protein spirals in pairs of the same structure as ds DNA to attract and hold ds DNA and provide synthesis and decomposition of different metabolites simultaneously. The conclusion of the polypeptide spiral of the proteins spiral is that helicoid ds DNA associated with helicoid polypeptide spirals can perform only protein and spiral nucleic acid biosynthesis and metabolism of different metabolites simultaneously.
- Polypeptide spirals that follow spirals of ss DNA of ds DNA are called protein spirals in this book.
- Two protein spirals with two inside polypeptide spirals with a ds DNA structure build the string. The protein spirals of the strings must move each other in opposite directions to provide synthesis and decomposition of metabolites and protein and nucleic acid spirals biosynthesis simultaneously. This conclusion is based on the fact that metabolism occurs among prokaryotic cells' proteins moving by each other.
- Strings must be organized in pairs of filaments to allow growth and division. Pairs of strings of specific organization build filaments.

Filaments with specific organization of strings allow division of filaments. Divisions of filaments allow binary fissions of prokaryotic cells.

- The model of filament has two strings with coiled and straight (linear) parts. The coils have growing ends of strings and grow around the linear parts of the strings (see picture above). Linear parts of the same strings with coils have nongrowing ends. Linear parts do not grow. The new model of filament is called the Rosalind Franklin model because her work allowed the building of the structure of ds DNA and the new model of the string. The model of the string enabled the building of the model of filament.

- Filaments are linear and interwoven. This conclusion ensued from the fact that electronic pictures of many prokaryotic cells show interwoven filaments.
- Polypeptide spirals of protein spirals of strings must attract each other via metabolites to make compact bodies of prokaryotic cells and allow binary fission.
- Strings must grow at one end and not at the other so filaments can grow and divide. The opposite poles of filaments must have the specific organization of strings previously described so filaments can grow at opposite poles and undergo binary fission.
- Protein spirals that move in opposite directions in the strings produce single protein spirals at the opposite ends. Single protein spirals that grow must get newly synthesized protein spirals so the string can grow and produce a single spiral at the opposite end. The single protein spiral at the nongrowing end must break the single protein spirals that the string does not grow. The growing end of the string must get protein spirals at its nongrowing end. The single protein spirals that are broken at the nongrowing end of the string must be used for movement of spirals of the nongrowing end of the other string. The strings at the nongrowing ends must interchange protein and nucleic acid spirals to grow. The single strings with this structure and property can provide growth of the cell but not their division.
- The glucose rings of NAM and NAG molecules have eight groups of specific atoms in eight vertical lines (chapter 3). These eight groups look for eight specific polypeptides with the same groups of atoms (metabolites) attracted by their specific coenzymes. The eight specific polypeptides with specific coenzymes and specific metabolites (groups of atoms) must exist to add the specific metabolites to glucose molecules of polysaccharide chains to make NAM and NAG of peptidoglycan chains.

(alfa - NH)
L - Alanine
D - Isoglutamine
L - Lysine or DAP
D - Alanine
(alfa - COOH)

Molecule of a glucose ring of NAM acid (NAM) or N-acetylglucosamine (NAG) has six carbon atoms. The four polypeptides with specific coenzymes attract carbon atoms of glucose rings.

- The eight polypeptides of different lengths with specific coenzymes and metabolites in different concentrations must attract each other along their lengths at the synthetic and the decomposition protein spirals to synthesize and decompose NAM and NAG and metabolize metabolites simultaneously. (See the structure of NAM and NAG in which each glucose ring of NAM and NAG brings different four-atom groups in four vertical lines.) The eight polypeptides must also bring different concentrations of specific coenzymes and metabolites to allow synthesis and decomposition of metabolites and different metabolism of different prokaryotic cells

polypeptide 1	polypeptide 2	polypeptide 3	polypeptide 4	polypeptide 5	polypeptide 6	polypeptide 7	polypeptide 8
ATP a. (--)	B$_1$, biotin, CoA c. (+)	B$_6$, Fol. e. (-)	NAD, FAD g.(+)	ATP i. (--)	B$_1$, biotin, CoA k. (+)	B$_6$, Fol. K m. (-)	NAD, FAD o. (+)
NAD, FAD b. (+)	K d. (-)	B$_{12}$ f. (+)	ATP h. (--)	NAD, FAD j. (+)	K, biotin l. (-)	B$_{12}$ n. (+)	ATP p. (--)

This table of specific polypeptides and coenzymes shows the attraction of the specific polypeptides in polypeptide spirals, the building of NAM acids and NAG, and the interaction of specific polypeptides of protein spirals during metabolism.

- Specific metabolites (appendix X) of specific coenzymes are electrically bipolar molecules. The metabolites with electrical charges exist in polypeptides to attract each other along their lengths to build polypeptide and protein spirals; this is possible because specific coenzymes are also electrically bipolar molecules.
- The two protein spirals of the strings must have specific quantitative relationships among metabolically connected coenzymes to provide decomposition and synthesis of the NAM, NAG, and other metabolites simultaneously. Protein spirals in the strings that collect more complex metabolites than simple ones are synthetic. The protein spirals in the strings that collect more simple metabolites than complex ones are the decomposition ones.
- Eukaryotic cells have many similarities with prokaryotic cells including the new structures and many similar phenomena. This conclusion ensued from comparison of new structures and different phenomena of prokaryotic and eukaryotic cells.

Understanding the new structures of prokaryotic and eukaryotic cells connected with metabolism and the phenomena was a slow process because it had to explain the phenomena, which are further analyzed and connected with other phenomena in the next chapters.

Chapter 3
Specific Polypeptides and Coenzymes

Models of specific polypeptides with specific coenzymes and metabolites[4] of prokaryotic cells are built by NAG and NAM acid molecules of peptidoglycan and different metabolites.

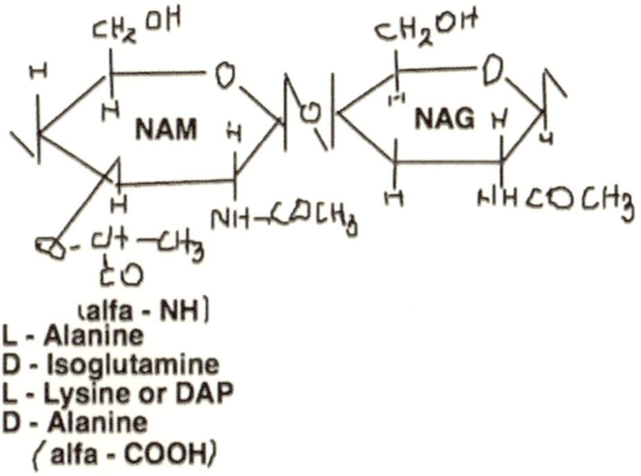
CH2OH
CH2OH
NAM
NAG
NH-COCH3
NHCOCH3
CO
(alfa - NH)
L - Alanine
D - Isoglutamine
L - Lysine or DAP
D - Alanine
(alfa - COOH)

Glucose rings of NAM acid (NAM) and N-acetylglucosamine (NAG) have six carbon atoms that attract specific groups of atoms.

Groups of atoms of NAM and NAG molecules are specific metabolites that exist in metabolism; they are connected by carbon atoms in the glucose rings.

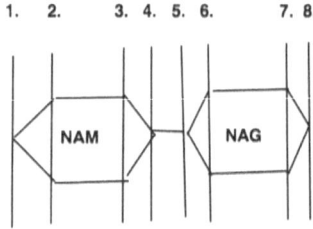

**The parallel lines cut the glucose rings of
NAM and NAG at their carbon atoms**

Glucose rings of NAG and NAM acid molecules have twelve groups of atoms connected with carbon atoms in their rings. The twelve specific groups of atoms can be grouped into eight groups according to the eight parallel lines that cut the glucose rings along their carbon atoms.

The glucose ring of NAM has four groups (see the above molecules of NAM and NAG). The first group has only a group of atoms (H-[C]-OH). The second group has three groups of atoms: (H – [C]-CH2OH), H –[C]-O –CH (CO) – CH3, and the specific polypeptides). The third group has a group (H –[C]-NHCOCH3). The fourth group has also a group (H-[C]-OH). The glucose ring of NAG also has four groups. The first group has only a group of atoms (H-[C]-OH). The second group has a group of atoms: (H –[C]-OH). The third group has a group (H –[C]-NHCOCH3). The fourth group has also a group (H-[C]-OH).

The eight polypeptides with specific coenzymes attract groups of atoms in the carbon atoms of glucose rings of NAM and NAG the same as the parallel lines of the glucose rings of NAM and NAG in the above scheme. The different metabolites are connected with specific coenzymes before putting specific coenzymes in the eight polypeptides. The metabolites and coenzymes are put in the eight specific polypeptides along their lengths according to the rules: metabolites attract each other by opposite electrical charges, specific coenzymes can add or take groups of atoms of NAM and NAG acid, and coenzymes can perform metabolism among the polypeptide spirals. The attractions among the specific metabolites along the lengths of specific polypeptides allow attractions of the eight specific polypeptides in the polypeptide spirals of protein spirals.

polypeptide 1	polypeptide 2	polypeptide 3	polypeptide 4	polypeptide 5	polypeptide 6	polypeptide 7	polypeptide 8
ATP a. (--)	B_1, biotin, CoA c. (+)	B_6, Fol. e. (-)	NAD, FAD g. (+)	ATP i. (--)	B_1, biotin, CoA k. (+)	B_6, Fol. K m. (-)	NAD, FAD o. (+)
NAD, FAD b. (+)	K d. (-)	B_{12} f. (+)	ATP h. (--)	NAD, FAD j. (+)	K, biotin l. (-)	B_{12} n. (+)	ATP p. (--)

Coenzymes as bipolar molecules have electrical charges that attract opposite electric charges of bipolar metabolites	Metabolites as bipolar molecules have electrical charges that attract opposite electric charges of bipolar coenzymes
ATP+	PO4--
FAD-, NAD-	H+
B_1+	pyruvate+
biotin+	different acids of Krebs cycle+
CoA	acetyl+
K vitamins	fatty acids-
B_6	amino acids-
folic acid	acetyl-
B_{12}	Fe++

The eight polypeptides must repeat along the polypeptide spirals to synthesize and decompose NAG and NAM molecules and other metabolites simultaneously. The protein spirals of strings of filaments of different prokaryotic cells with peptidoglycan molecules must have the same structure and position of specific coenzymes along specific polypeptides to produce peptidoglycan.

Prokaryotic cells have protein spirals with polypeptide spirals that move in opposite directions by different concentration of specific coenzymes. Protein spirals allow decomposition and synthesis of metabolites and movement simultaneously.

The first specific polypeptides of the decomposition and the synthetic protein spirals have ATP that attract negative PO4--ions and FMN that attract positive H+ ions.

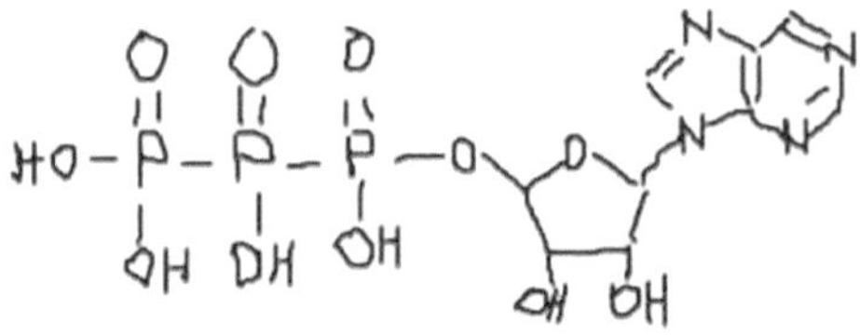

ATP

ATP coenzymes are on one side of the first polypeptides whereas FMN coenzymes are on the opposite side and form bipolar polypeptides. The first and eighth polypeptides together with the fourth and fifth polypeptides fuse glucose molecules in the polysaccharide chain in the synthetic protein spirals or they split the polysaccharide chains into fructose molecules and other simple metabolites in the decomposition protein spirals. The fused glucose molecules transform into NAM and NAG with the help of other polypeptides with specific coenzymes and metabolites. The first and fifth polypeptides have the same coenzymes (ATP and FMN) and their different concentrations. The eighth and fourth have the same coenzymes (ATP and NAD) and their different concentrations.

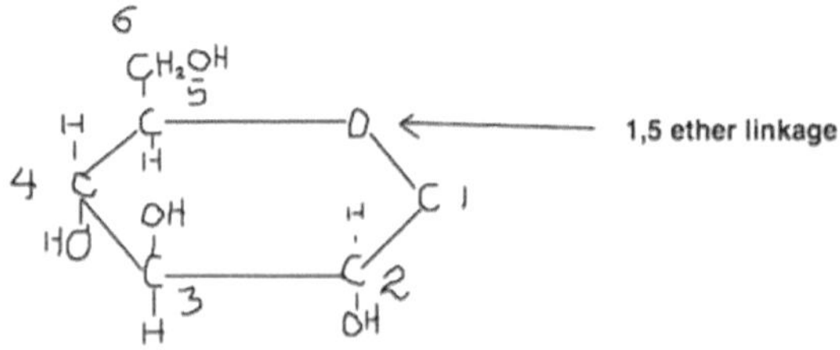

Glucose molecule

Flavin adenindinucleotide coenzymes (FAD) attract hydrogen ions from NAD coenzymes during polarization.

FAD

The second specific polypeptides of decomposition and synthetic protein spirals have B_1, biotin, CoA, and K vitamins. B_1, biotin, and CoA as electrically bipolar molecules attract negative carboxyl groups of electrically bipolar acids and free their positive methyl groups. K vitamins attract the chains and the positive methyl groups of the electrically bipolar fatty acids and free their negative carboxyl groups at opposite ends. B_1, biotin, and CoA are on one side of the second specific polypeptide whereas K vitamins are on the opposite side. B_1 vitamins attract pyruvate molecules from glyceraldehide-3P attracted by ATP. Biotins attract metabolites from the Krebs cycle. CoA attracts acetate as acetyl-CoA from folic acids and K vitamins that attract fatty acids from fats and CoA.

B_1 vitamins (coenzymes) of the second polypeptides fuse pyruvate molecules to C3 of glucose molecules that are the precursors of NAM. This is possible only if B_1 vitamins have accumulated pyruvate molecules during polarization. B_1 will have accumulated pyruvate molecules only if ATP coenzymes of close polypeptides take fewer pyruvate molecules from B_1 during polarization. This is different for the sixth polypeptides with B_1 involved in making NAG.

B_1 of the sixth polypeptide does not accumulate enough pyruvate molecules during polarization because ATP coenzymes of close polypeptides take much more pyruvate molecules from them. B_1 vitamins of the sixth polypeptides do not have enough pyruvate molecules to fuse them to the C3 of glucose molecules, the precursors of NAG. This concept will be more understandable after an explanation of decomposition and synthesis of glucose molecules and the coenzymes involved in both opposite processes. In summary, B_1 vitamins are involved in the decomposition and the synthesis of glucose molecules and synthesis of NAM and NAG.

Thiamine pyrophosphate TPP

Vitamins B_1 attract pyruvate molecules to their carboxyl groups.

Biotin vitamins are involved in transformation of different metabolites present during the Krebs cycle. Biotins attract carboxyl groups that are products of metabolism of metabolites. Biotins are in connections with folic acids, B_6, and B_{12} that belong to the third or seventh specific polypeptides.

Biotin

Biotin coenzymes attract the metabolites from Krebs cycle for their carboxyl groups during polarization.[61]

CoA coenzymes attract acetyl groups of fatty acids and expose their methyl groups that bring positive electrical charges. Biosynthesis and decomposition of fatty acids take place in CoA. They are in connection with folic acids that transfer acetyl molecules from or to them and K vitamins that transfer fatty acids to or from them.

Functions of CoA are not clearly explained in the literature.

CoA is involved in metabolism of NAM. CoA attracts most likely the specific pentapeptides that belong to NAM acids. They are L-alanine, D-isoglutamine, L-lysine or diaminopimelic acids, and D-alanine. CoA connects NAM and NAG chains in the net of peptidoglycan.

K vitamins likely attract fatty acids by their long chains and expose their methyl groups with the positive electrical charges (hypothesis).

Functions of K vitamins in the metabolism of carbohydrates and fatty acids are not clear in the literature.

Coenzymes and metabolites of the second specific polypeptides connect with metabolites and specific coenzymes of the second and fourth polypeptides of the same and close polypeptide spirals.

The third specific polypeptides of decomposition and synthetic protein spirals have B_6, folic acids, and B_{12}. B_6 and folic acids with their metabolites are on one side of the second specific polypeptides whereas B_{12} is on their other side. B_6 attract amino groups and other metabolites. The molecules attracted by B_6 as bipolar molecules have free carboxyl groups. Folic acids attract methyl groups of acids and metabolites. Molecules attracted by folic acids as bipolar molecules have free carboxyl groups. B_{12} attracts $Fe++$ ions that attract porphobilonogens from biotins during depolarization.

B_6 coenzymes are involved in metabolism of amino acids and other metabolites.

B_6 is also involved in decomposition and synthesis of glucose metabolites, fatty acids, and amino acids with other coenzymes. Amino acids are bipolar molecules as are acetyl and fatty acids. B_6 coenzymes attract amino acid molecules for their amino (–NH2) groups and expose their carboxyl groups that bring negative electrical charges.

B_6 coenzymes fuse amino groups with the help of NAD coenzymes to C2 of glucose molecules of the precursors of NAM or NAG. Folic acids from the third and seventh polypeptides transfer acetate groups of metabolites to (–NH-) that are already fused to C2 of glucose molecules and make the precursors of NAM or NAG.

Folic acids are involved in decomposition and synthesis of pyruvate during polarization according to the new hypothesis. Folic acids transfer acetate molecules to the metabolites of B_1, CoA coenzymes, biotins, amino groups attracted by C2 of the precursors of NAM and NAG, and other metabolites, or they take acetates from their metabolites.

Metabolites attracted by B_1, CoA, biotin coenzymes, NAM, and NAG transfer acetates to folic acids with the help of other coenzymes. This logic will be clear after explanation of the synthetic and decomposition protein spirals and their coenzymes that enable decomposition and synthesis of glucose molecules, fatty acids, amino acids, and other molecules.

Folic acids attract acetates for their-CH3 groups and transfer them to the different metabolites that are attracted by biotin, CoA, and B_6 or take acetates from them.

B_{12} vitamins are initiators of synthesis of porphyrin-Fe++ complexes involved in oxidation of H+ ions according to the new hypothesis. The unknown coenzymes (B_{12}?) or amino acids of the specific polypeptides release Fe++ ions during depolarization. B_{12} attracts porphobilinogens connected with biotins into porphyrin molecules that attract Fe++ and make the porphyrin-Fe++ complex. This occurs during depolarization.

The porphyrin ring with Cobalt ions

Dimethylbenzimidazole

B_{12} attracts porphobilinogen into porphyrin molecules that attract Fe^{++} ions during depolarization.

The fourth specific polypeptides have ATP that attract negative $PO4^{--}$ions at an end of polypeptides and NAD that attracts H^+ ions at their opposite ends. The fourth and fifth polypeptides that have ATP and FMN coenzymes with specific ions fuse glucose molecules. The fused glucose molecules of polysaccharide chains transform later into NAM and NAG of peptidoglycan chains that alternatively repeat in them.

NAD coenzymes are involved in dehydrogenation of molecules during polarization. NAD coenzymes transfer hydrogen atoms as ions to FAD coenzymes of the fifth polypeptides. The first, fourth, fifth, and eighth polypeptides have different concentrations of ATP.

The fifth specific polypeptides (similar to the first polypeptides) have ATP that attracts negative $PO4^{--}$ions at an end of polypeptides and FMN that attracts H^+ ions at opposite ends. The fifth and fourth polypeptides with ATP and NAD coenzymes with specific ions fuse glucose molecules into polysaccharide or split polysaccharide into glucose molecules. The fused glucose molecules transform into NAM and NAG that alternatively repeat at peptidoglycan molecules.

NAD coenzymes of the fourth polypeptides transfer H+ ions to FAD coenzymes of the fifth polypeptides during polarization. FAD coenzymes release hydrogen atoms during depolarization. ATP coenzymes attract hydroxyl groups of different molecules and are involved in decomposition and biosynthesis of glucose and the other molecules.

The sixth specific polypeptides of decomposition and synthetic protein spirals have also B_1, biotin, CoA, and K vitamins. The sixth and the second polypeptides have the same coenzymes but different concentrations. B_1, biotin, and CoA as electrically bipolar molecules attract negative carboxyl groups of electrically bipolar acids and free their positive methyl and other groups. K vitamins attract the chains and the positive methyl groups of electrically bipolar fatty acids and free their negative carboxyl groups at opposite ends. B_1, biotin, and CoA are on one side of the second polypeptides whereas K vitamins are on the opposite side. B_1 attracts pyruvate molecules from glyceraldehide-3P attracted by ATP. Biotins attract specific metabolites from the Krebs cycle. CoA attracts acetate as acetyl-CoA from folic acids and K vitamins that attract fatty acids from the fats and CoA.

B_1 vitamins of the sixth polypeptides do not fuse the pyruvate molecules to C3 of glucose molecules that are precursors of NAG. B_1 vitamins do not accumulate pyruvate molecules during polarization. B_1 will not have accumulated pyruvate molecules only if ATP coenzymes of close polypeptides take more pyruvate molecules from B_1 during polarization. This is different at the second polypeptides with B_1 that are involved in making NAM.

B_1 of the second polypeptides accumulate enough pyruvate molecules during polarization because ATP coenzymes of close polypeptides take fewer pyruvate molecules from them. This concept will be more understandable after an explanation of decomposition and synthesis of glucose molecules and the coenzymes involved in both opposite processes.

Thiamine pyrophosphate TPP

Vitamin B_1 attracts pyruvate molecules to their carboxyl groups.

Biotin vitamins are involved in transformation of metabolites during the Krebs cycle. Biotins attract carboxyl groups that are products of the metabolism of metabolites. Biotins are in connections with folic acids, B_6, and B_{12} that belong to the third or seventh specific polypeptides.

Biotin

Biotin coenzymes attract the metabolites from the Krebs cycle for their carboxyl groups during polarization.[61]

CoA coenzymes attract acetyl groups of fatty acids and expose their methyl groups that bring positive electrical charges. Biosynthesis and decomposition of fatty acids take place in CoA. They are in connection with folic acids that transfer acetyl molecules from or to them and K vitamins that transfer fatty acids to or from them.

The functions of CoA are not clearly explained in the literature.

CoA is involved in the metabolism of NAM. CoA attracts most likely the specific pentapeptides that belong to NAM acids: L-alanine, D-isoglutamine, L-lysine or diaminopimelic acids, and D-alanine. CoA connects the chains of NAM and NAG in the net of peptidoglycan.

K vitamins attract fatty acids by their long chains and expose their methyl groups with positive electrical charges (hypothesis).

Functions of K vitamins in the metabolism of carbohydrates and fatty acids is not clear in the literature.

Coenzymes and metabolites of the sixth specific polypeptides connect with metabolites and coenzymes of the fifth and seventh specific polypeptides of the same and close polypeptide spirals.

The seventh specific polypeptides of decomposition and synthetic protein spirals have B_6, folic acids, and B_{12} vitamins. B_6 and folic acids with their metabolites are on one side of the seventh specific polypeptides whereas B_{12} is on the other side. B_6 attracts amino groups of amino acids and other metabolites. The molecules attracted by B_6 as bipolar molecules have free carboxyl groups. Folic acids attract methyl groups of the different acids and the other metabolites. Molecules attracted by folic acids as bipolar molecules have free carboxyl groups. B_{12} attracts Fe++ ions that attract porphobilonogens from biotins during depolarization.

B_6 coenzymes are involved in the metabolism of amino acids and other metabolites.

Pyridoxal - 5'- phosphate (PLP)   Pyridoxine (Vitamin B6)

B_6 is also involved in decomposition and synthesis of glucose metabolites, fatty acids, and amino acids with other coenzymes. Amino acids are bipolar molecules as are acetyl and fatty acids. B_6 coenzymes attract amino acids molecules for their amino (–NH2) groups and expose their carboxyl groups that bring negative electrical charges.

B_6 coenzymes fuse amino groups by the help of NAD coenzymes to C2 of glucose molecules of the precursors of NAM or NAG. Folic acids from the third and seventh polypeptides transfer acetate groups of the metabolites to –NH-that are already fused to the C2 of glucose molecules and make the precursors of NAM or NAG.

Folic acids are involved in decomposition and synthesis of pyruvate during polarization according to the new hypothesis. Folic acids transfer acetate molecules to the metabolites of B_1, CoA coenzymes, biotins, the amino groups attracted by C2 of the precursors of NAM and NAG and other metabolites or take acetates from their metabolites. The metabolites attracted by B_1, CoA, biotin coenzymes, NAM, and NAG transfer acetates to folic acids with the help of other coenzymes. This logic will be clear after explanation of the synthetic and decomposition protein spirals and their coenzymes that enable decomposition and synthesis of glucose molecules, fatty acids, amino acids, and other molecules.

Folic acids attract acetates for their-CH3 groups and transfer them to the different metabolites that are attracted by biotin, CoA, and B_6 or take acetates from them.

B_{12} vitamins are initiators of synthesis of porphyrin –Fe++ complexes that are involved in the oxidation of H+ ions according to the new hypothesis. The unknown coenzymes (B_{12}?) or amino acids of the specific polypeptides release Fe++ ions during depolarization.

B_{12} attracts porphobilinogens connected with biotins into porphyrin molecules that attract Fe++ and make porphyrin −Fe++ complex. This occurs during depolarization.

The porphyrin ring with Cobalt ions

Dimethylbenzimidazole

B_{12} attracts porphobilinogen into porphyrin molecules that attract Fe++ ions during depolarization.

The eighth specific polypeptides (similar to the fourth polypeptides) have ATP that attract negative PO4--ions at an end of polypeptides and NAD that attracts positive H+ ions at opposite ends. The eighth and the first polypeptides that have ATP and FMN coenzymes with their specific ions fuse glucose molecules. The fused glucose molecules of polysaccharide chains transform later into NAM and NAG of peptidoglycan chains that alternatively repeat in them.

NAD coenzymes are involved in dehydrogenations of the different molecules during polarization. NAD coenzymes transfer hydrogen atoms as the ions to FMN coenzymes of the fifth polypeptides. The first, fourth, fifth, and eighth polypeptides have different concentrations of ATP.

The eight polypeptides with metabolites are also present in the strings of eukaryotic cells. The polypeptides with metabolites of eukaryotic cells are much longer than those of prokaryotic cells. They have nonprotein hormones with ions that make better attraction among them. The polypeptides are described in the part of eukaryotic cells.

Specific Polypeptides and New Protein Structures

Polypeptides attract each other to build the strings' two protein spirals. One spiral is synthetic whereas the other is involved in decomposition. They have the same polypeptides with different concentrations of the same coenzymes, which allow biosynthesis and decomposition of glucose, polysaccharides, fatty acids, fats, amino acids, nucleotides, coenzymes and their movement in opposite directions simultaneously.

The porphyrin ring with Cobalt ions

Dimethylbenzimidazole

Protein spirals consist of polypeptide spirals, which consist of polypeptides attracted along the lengths.

Specific coenzymes attract specific metabolites during polarization and release them during depolarization. Polarization and depolarization alternate during metabolism. Polarization allows movement of protein spirals whereas depolarization allows neutralization of hydrogen atoms, metabolism's major waste products.

Simultaneous metabolism of carbohydrates, fatty acids, fats, amino acids, and the other molecules, and simultaneous existence of the different phenomena of prokaryotic and eukaryotic cells are possible only if specific coenzymes attract specific metabolites (facts form the literature prove this) and if an organization of specific coenzymes exists in specific polypeptides organized at the specific protein structures (facts that the different phenomena take places at the same time prove an existence of specific polypeptides and specific protein structures that enable this).

The synthetic and decomposition protein spirals and polarization and depolarization will be discussed in detail in other chapters. Before that, I will explain metabolism carbohydrates, fatty acids, fats, amino acids, and the neutralization of hydrogen atoms according to the new approach due to better understanding of inseparability of metabolism of the different molecules, the structures, and the different phenomena.

CHAPTER 4
COENZYMES OF POLYPEPTIDES AND CARBOHYDRATES

Coenzymes work together during metabolism of metabolites (carbohydrates, fatty acids, amino acids, and others) according to strict metabolic relationships among metabolites during decomposition and synthesis. Some coenzymes attract metabolites from others attracted by coenzymes. This occurs only if some specific coenzymes have bigger concentrations than do others, which allows greater attraction forces.

Many enzymes are involved in synthesis and decomposition of metabolites in prokaryotic and eukaryotic cells. Enzymes consists of protein, coenzymes, and metabolites after catalytic reactions. Enzymes involved in decomposition and synthesis are not mentioned here due to easier explanations of metabolism. The same coenzymes make different enzymes that can attract complex molecules depending on the quality and quantity of their proteins. Coenzymes with specific metabolites of enzymes always attract specific groups of atoms of those metabolites (appendix XI).

ATP, NAD, and FAD coenzymes of the eighth, first, fourth, and fifth polypeptides of decomposition protein spirals attract, break, and transform glucose molecules into metabolites during glycolysis.

ATP coenzymes attract hydroxyl groups of polysaccharide molecules and break polysaccharide chains into free and incomplete molecules of glucose by adding energy to them. (See the different polysaccharides on the next page.) Free and incomplete molecules of glucose attract hydrogen atoms released during depolarization and complete molecules immediately after breaking the polysaccharide chains. ATP coenzymes of the eighth, first, fourth, and fifth polypeptides of decomposition and synthetic protein spirals attract complete glucose molecules for their hydroxyl groups of C1 or C6 atoms.

Different polysaccharides look for different enzymes that have the same coenzymes for their disassembling.

Movements of protein spirals by each other during metabolism allow metabolism. Movement of polypeptides of new structures allow interaction of coenzymes during metabolism. Glycolysis of glucose molecules and their metabolites proves movement of specific polypeptides in polypeptide and protein spirals. ATP coenzymes of the first, fourth, fifth, and eighth polypeptides of the decomposition polypeptide spirals attract glucose 6 phosphate molecules and transform them among themselves into glucose 6,1 phosphate and later in fructose 1,6 phosphate.

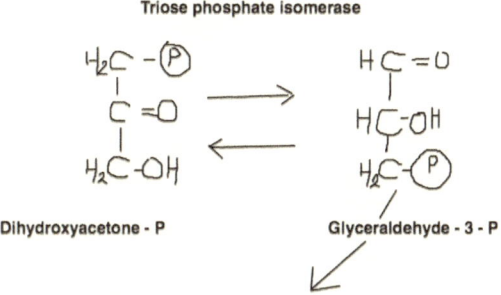

Polypeptides of decomposition protein spirals move and split fructose 1,6 phosphate among their polypeptides in dihydroxyacetone phosphate and glyceraldehyde-3 phosphate, which transform into each other during metabolism.

They transform into 1,3-biphosphoglycerate, which transforms into 3-phosphoglycerate that transforms further into 2-phosphoglycerate. This transformation can take place only by movement of polypeptides of the same and close protein spirals and involvement of different ATP of close polypeptides of decomposition protein spirals.

37

The second and sixth polypeptides have B_1, biotin, CoA, and K vitamins. Their B_1 attracts pyruvate from phosphoenolpyruvate of the first and fifth polypeptides of close polypeptide spirals of decomposition protein spirals. Folic acids of the third and seventh polypeptides of close polypeptide spirals attract acetate ions from pyruvates attracted by B_1. B_1 attracts COO-group after the splitting of pyruvates and releases COO-group as CO2 during depolarization. Folic acids transfer acetate to CoA coenzymes attracted by the second and sixth polypeptides of close polypeptide spirals of decomposition protein spirals. Folic acids transfer acetate groups to oxaloacetate attracted by biotin of close polypeptide spirals of the decomposition protein spirals.

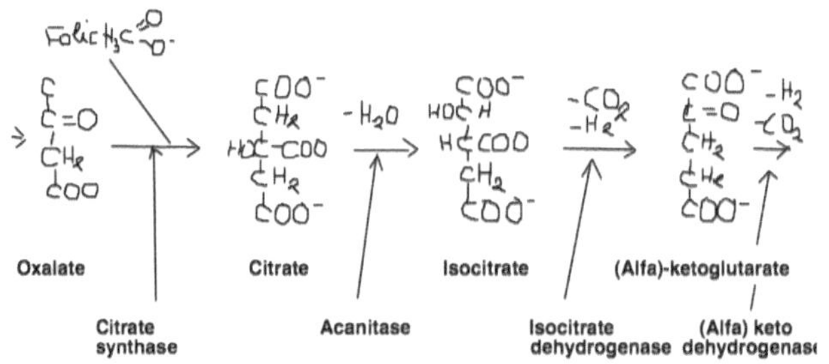

Acetate groups with oxaloacetate fuse into citrate molecules attracted by biotins, which transform them into molecules of the Krebs cycle by specific coenzymes of close polypeptide spirals of decomposition protein spirals.

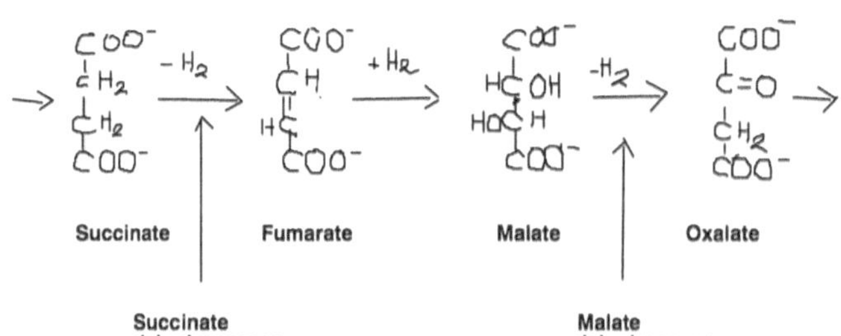

Acetyl-CoA transfers acetyl to oxaloacetate attracted by biotin according to the literature without any explanation of why, how, or where. (See the Krebs cycle). If CoA coenzymes transfer acetyl

(CH3COO-CoA) to oxaloacetate, CoA cannot be used for the biosynthesis of fatty acids because (COO-) group of acetyl-CoA must be free (not attracted) for their biosynthesis. This contradiction can be explained only if folic acids transfer acetate molecules by their methyl groups (folic acid CH3 COO-) to different metabolites attracted by different coenzymes (CoA, biotin, B_6) instead of CoA according to the new view. CoA attracts acetate molecules for their COO-groups as acetyl-CoA also according the new view. They are used for biosynthesis and decomposition of fatty acids also according to the new view.

Citrate molecules in biotins under influence of NAD and ATP coenzymes transform into isocitrate molecules. (See the Krebs cycle according to the literature.) The Krebs cycle[5] is changed to satisfy logic among metabolically connected coenzymes according to the new approach.

Isocitrate molecules in biotins are attracted by NAD and biotin coenzymes of close protein spirals. NAD coenzymes take hydrogen atoms from C2 atoms of isocitrates whereas biotin coenzymes take carboxyl groups from their C3 atoms. They transform isocitrates into alpha-ketoglutarate. Alpha-ketoglutarates in biotins are attracted by biotins coenzymes of close polypeptide spirals. Biotin coenzymes take carboxyl groups of alpha-ketoglutarates and transform them into succinates, which are attracted by close and empty CoA coenzymes that transform them into succinyl-CoA. Succinyl-CoA attracts glycine from B_6 of close protein spirals and transforms them into delta-aminolevulinate, a precursor of porphobilinogen. Succinates in biotins not attracted by CoA coenzymes are attracted by NAD coenzymes of close polypeptide spirals that transform them into fumarates, which occurs during polarization. Fumarates in biotins transform into malonates during depolarization under the influence of released hydrogen ions from FMN and hydroxyl ions from ATP coenzymes.

Hydrogen atoms of malonates in biotin coenzymes are attracted by NAD coenzymes during polarization and transform malonates into oxaloacetates, and oxaloacetates in biotin attract acetates from folic acids, and with this, the Krebs cycle of the different metabolites starts again. This enables the synthesis of the precursors of glucose molecules and gluconeogenesis in the synthetic protein spirals.

Gluconeogenesis is a process opposite of glycolysis. Malate molecules attracted by biotins of the synthetic protein spirals change in fumarate also in biotin coenzymes under influence of NAD coenzymes during polarization. Fumarate molecules change in succinate molecules also in biotins of the synthetic protein spirals during depolarization.

Biotins with carboxyl groups of close polypeptide spirals transfer carboxyl groups to succinate molecules that transform into alpha-ketoglutarate molecules also in biotin coenzymes of the synthetic protein spirals. Alpha-ketoglutarate molecules are freed during depolarization and transform into isocitrate molecules. Biotin coenzymes of the synthetic protein spirals attract isocitrate molecules and transform them into citrate molecules.

Folic acid coenzymes of the synthetic protein spirals attract acetate molecules from citrate in biotin coenzymes and transfer them to carboxyl groups of B_1 coenzymes of the synthetic protein spirals. B_1 coenzymes with carboxyl groups fuse them into pyruvate molecules. ATP coenzymes of the synthetic protein spirals attract pyruvate molecules and transform them into phosphoenolpyruvate molecules, which are attracted by ATP and transform into 2-phosphoglycerate during depolarization.

ATP coenzymes of synthetic protein spirals attract 2-phosphoglycerate molecules that transform into 3-phosphoglycerate ones. Other ATP of the synthetic protein spirals attract 3-phosphoglycerate molecules and transform them into 1,3 biphosphoglycerate molecules, precursors of glyceraldehyde-3 phosphate and dihydroxyacetone phosphate, which fuse into glucose molecules among synthetic protein spirals.

ATP, NAD, and FAD coenzymes of the eighth first, fourth, and fifth polypeptides of synthetic protein spirals fuse glucose molecules in a chain.

1,5 ether linkage

ATP coenzymes of the first and fifth polypeptides attract hydroxyl groups from the fourth carbon atoms of glucose molecules, whereas NAD coenzymes of the fourth and eighth polypeptides attract hydrogen atoms from the first carbon atoms of glucose molecules simultaneously during polarization. Glucose molecules without hydrogen atoms at its first carbon atoms and without hydroxyl groups at its fourth carbon atoms can attract and fuse each other into polysaccharide chains during polarization.

The string consists of synthetic and decomposition protein spirals. Decomposition of polysaccharides, glucose, and their metabolites occurs in decomposition and synthetic protein spirals. Synthesis of metabolites of glucose molecules, glucose, and polysaccharides occurs also in the synthetic and the decomposition protein spirals. The spirals decompose and synthesize glucose molecules and their metabolites simultaneously. This approach will be clear after explanation of the interaction among coenzymes among protein spirals and their movements in chapter 8.

NAD coenzymes transfer hydrogen ions to FAD coenzymes that accumulate them also during polarization. NAD and FAD coenzymes are functionally connected during the transfer of electrons during oxidation according to the contemporary literature (appendix XI). I think NAD and FAD are involved in transferring hydrogen atoms as ions instead of electrons during polarization and depolarization of cells. NAD coenzymes attract and take hydrogen atoms from metabolites and transfer them to FAD coenzymes, which collect hydrogen ions up to the saturation point.

The amount of collected hydrogen ions at FAD depends on concentrations of FAD along the first polypeptides in protein spirals. When FAD coenzymes cannot attract any more hydrogen ions from NAD, they release them and start depolarization. They release hydrogen atoms because FAD coenzymes cannot attract them due to lack of attraction energy and because NAD coenzymes continuously transfer hydrogen ions to FAD coenzymes weakening and breaking their attraction energy.

These coenzymes relationships exist among coenzymes and their metabolites among the polypeptides of decomposition and synthetic protein spirals simultaneously during decomposition and biosynthesis of glucose molecules and polysaccharides in this way.

ATP-> ATP-> B_1-> folic acid-> biotin->biotin
ATP <-ATP <-B_1 <-folic acid <-biotin <-biotin

Synthesis and Decomposition of NAG and NAM Acids and Peptidoglycan

Synthesis of NAG and NAM and peptidoglycan occurs on glucose molecules of polysaccharides. ATP from the first or fifth polypeptides attracts hydroxyl groups of the third carbon (C3) of glucose molecules. B_1 vitamins add pyruvate (CH3COCO) to the third carbon of glucose molecules whereas CoA of the second or sixth polypeptides from close protein spirals adds specific polypeptides of the five amino acids (L-alanine, D-glutamine, L-glutamine, D-alanine, D-alanine) to the pyruvate (CH3COCO) of the third carbon of glucose molecules. Why is CoA chosen?

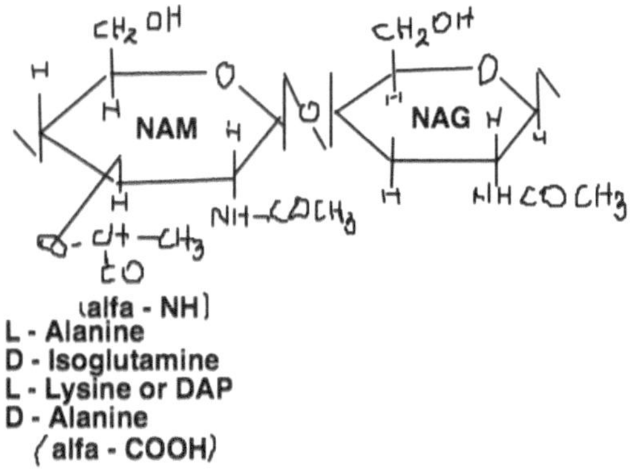

(alfa - NH)
L - Alanine
D - Isoglutamine
L - Lysine or DAP
D - Alanine
(alfa - COOH)

CoA coenzymes attract probably specific polypeptides produced by specific genes during biosynthesis of polypeptides, or their chain attracts probably five specific amino acids and fuse them into polypeptides. The logic of this hypothesis is that polypeptides must be produced and transferred by unknown molecules close to specific coenzymes in polypeptides of polypeptide spirals during biosynthesis of NAM and NAG. CoA was chosen due to its long chain and the structure that could attract specific polypeptides. The long molecules of CoA and its structure could also connect NAM acids of peptidoglycan chains by polypeptides in the mesh of peptidoglycan.

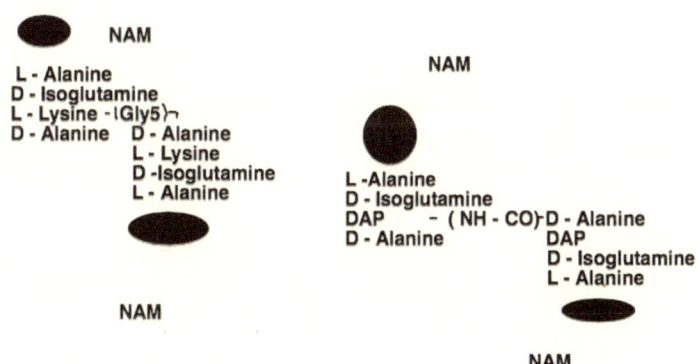

CoA attracts specific polypeptides and connects by them NAM acids of the peptidoglycan chains according to the hypothesis.

CoA coenzymes with their long chains probably attract specific amino acids from B_6 coenzymes and form polypeptides (L-alanine, D-isoglutamine, L-lysine, DAP, D-alanine, D-alanine) on their chains. The new hypothesis looks for proof not found in the literature.

The long CoA coenzymes with the last amino acid D-alanine of specific polypeptides attract probably also L-lysine or DAP of close specific polypeptides also attracted by other CoA. The polypeptides attract chains with NAM acids into nets of huge molecules of peptidoglycan. See the peptidoglycan molecule on the next page.

NAM

L - Alanine
D - Isoglutamine
L - Lysine -(Gly5)₇
D - Alanine   D - Alanine
L - Lysine
D -Isoglutamine
L - Alanine

NAM

NAM

L -Alanine
D - Isoglutamine
DAP   - (NH - CO)-D - Alanine
D - Alanine   DAP
D - Isoglutamine
L - Alanine

NAM

Different connections between the short polypeptides of NAM acids into peptidoglycan

Synthesis and decomposition of peptidoglycan molecules take place in filaments also simultaneously. Peptidoglycans have protective and energy storage functions of prokaryotic cells. Synthesis and decomposition of peptidoglycan molecules are products of specific polypeptides, coenzymes, and metabolites.

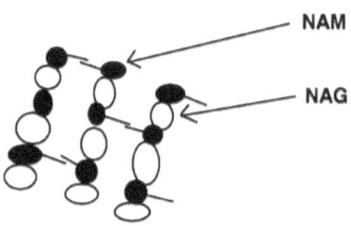

Chains with NAG and NAM acid are connected in the net of the huge peptidoglycan molecules at their NAM acids by specific polypeptides.

The main function of CoA coenzymes of synthetic protein spirals is to attract acetate molecules from folic acid coenzymes and synthesize fatty acids by folic acid, NAD, and ATP coenzymes according to the new view. CoA of decomposition protein spirals has fatty acids decomposed in acetate molecules by folic acid, NAD, and ATP coenzymes also according to the new view. The new view ensues from the fact that acetate metabolites are bipolar molecules and that COO-groups as the electrically negative side of acetate molecules are connected for CoA. Methyl groups of acetate molecules as the electrically positive side are connected by folic acids that transfer acetates to CoA. I modified the existing view of biosynthesis (anabolism) and decomposition (catabolism) of fats by this logical approach.

CHAPTER 5
COENZYMES OF POLYPEPTIDES AND FATS

The existing view of biosynthesis (anabolism) and decomposition (catabolism) of fats is modified according to the new approach,[6] which ensued from the fact that acetates and fatty acids are bipolar molecules. CoA of polypeptides attracts COO-groups of acetate molecules and fatty acids whereas folic acids of the polypeptides attract methyl groups of acetate molecules and fatty acids according to the new approach.

K vitamins attract fatty acids by similar chains according to the new approach. CoA, folic acid, K vitamins, NAD, and ATP of close polypeptides are involved in biosynthesis and decomposition of fatty acids and fats that takes place in polypeptides of synthetic and decomposition protein spirals simultaneously.

Biosynthesis of fatty acids occurs through several phases via malonic acids according to the literature. The phases are attachment, condensation, reduction, dehydration, and reduction.

The attachment phase starts when acyl carrier proteins (ACP) attract and take acetyl ions from acetyl-CoA and transform it into acetyl-ACP and CoA by ACP transacylase. ACP attracts and takes malonyl ions from malonyl-CoA and transform it into malonyl-ACP and CoA by ACP transacylase.

The condensation phase starts when acetyl-ACP reacts with chain-extending malonyl-ACP by 3-ketoacyl synthase.

The reduction phase starts when 3-ketoacyl reductase reduces the chain-extending malonyl-ACP.

The dehydration phase starts when the reduced chain-extending malonyl-ACP loses water by 3-hydroxylacyl ACP dehydrase.

The reduction phase starts when the last compound is reduced in the fatty acids by enoil-ACP reductase.

Synthesis of saturated fatty acids by fatty acid synthase II in *E. coli* according to the literature.[6]

The literature explains biosynthesis and decomposition of fatty acids in the phases by enzymes without any connections with structures of the cells due to reasons offered in the introduction, so decomposition of fatty acids according to the literature will not be discussed here. Biosynthesis and decomposition of fatty acids according to the new approach are explained differently because these processes are connected with cell structures.

New Approach to Biosynthesis of Fatty Acids and Fats

Biosynthesis and decomposition of fatty acids and fats occurs in protein spirals of strings simultaneously. Biosynthesis occurs in synthetic protein spirals whereas decomposition occurs in decomposition protein spirals. Metabolites of synthetic and decomposition spirals are involved in biosynthesis and decomposition.

Folic coenzymes of polypeptides attract methyl groups of acetate molecules as electrically positive parts of bipolar acetate molecules. Folic acids transfer acetate to CoA of close specific polypeptides, which attracts carboxyl groups of acetate molecules as electrically negative parts of the bipolar acetate molecules. CoA transfers acetate to folic acids of close specific polypeptides.

Empty CoA coenzymes of close protein spirals attract acetate molecules from folic acids for their carboxyl group in the complex molecule acetyl-CoA during polarization (the attachment phase according to the literature). Polarization and depolarization change alternatively during metabolism of prokaryotic and eukaryotic cells and will be explained later in greater detail.

The methyl groups (-CH3) of acetyl-CoA attract –OOC-groups of acetates attracted by folic acids of closest protein spirals or the same close polypeptide spirals. ATP and NAD coenzymes take OH-and H+ ions (dehydration) of acetate molecules and condense the molecules into CoA-OOC-CH2-OC-CH3 molecules during polarization (condensation phase according to the literature). Reductions of CoA-OOC-CH2-OC-CH3 molecules in CoA-OOC-CH2-HCOH-CH3 molecules take place during depolarization. Reduction here means adding hydrogen to molecules. ATP and NAD coenzymes take OH- and H+ from the CoA-OOC-CH2-HCOH-CH3 molecules and transform them into CoA-OOC-CH = HC-CH3 during polarization. CoA-OOC-CH = HC-CH3 gets hydrogen ions during depolarization and transform into CoA-OOC-CH2-CH2-CH3 (fatty acids).

The new cycle of biosynthesis of molecules starts when CoA-OOC-CH2-CH2-CH3 attract –OOC-groups of acetates attracted by folic acids of close protein spirals or the same close polypeptide spirals. ATP and NAD coenzymes take OH-and H+ ions of the attracted groups of acetate and CoA-OOC-CH2-CH2-CH3 molecules and condense them in CoA-OOC-CH2-CH2 – CH2-OC-CH3 molecules during polarization according to the new view.

The reduction of CoA-OOC-CH2-CH2 – CH2-OC-CH3 in CoA-OOC-CH2-CH2 – CH2-HCOH-CH3 molecules occurs during depolarization. The ATP and NAD coenzymes take OH-and H+ of the CoA-OOC-CH2-CH2 – CH2-HCOH-CH3 molecules and transform them into CoA-OOC-CH2-CH2-CH = HC-CH3 during polarization. The CoA–OOC-CH2-CH2-CH = HC-CH3 get hydrogen ions during depolarization and transform into CoA-OOC-CH2-CH2-CH2-CH2-CH3. The polarization and depolarization allow attractions, transformation, and growth of fatty-acid chains. The described cycle allows growth of fatty acids.

K vitamins probably bind fatty acids from CoA and transfer them to 3-phoshoglicerate molecules attracted by ATP (the new hypothesis). Fatty acids and 3-phoshoglicerate make fats in ATP by their fusion with the help of NAD and ATP coenzymes of close specific polypeptides.

Functions of K vitamins in the metabolism of carbohydrates and fatty acids are not clear in the literature.

K vitamins transfer fatty acids because they have a long chain that resembles a fatty acid. K vitamins with repeating molecular groups easily bind long molecules of fatty acids according to the new approach. K vitamins as fat-soluble coenzymes could be involved in the transfer of fatty acids from CoA to 3-phoshoglycerate and vice versa, from 3-phosphoglycerate to CoA. This new hypothesis needs to be proven. They are put in the third and seventh polypeptides with CoA because they are soluble in fats.

New Approach to Decomposition of Fatty Acids and Fats

Beta-oxidation of fatty acids (decomposition) is the opposite of their biosynthesis. Beta-oxidation of fatty acids has the same phases according to the literature. The phases of the literature are transfer of fatty acids, beta-oxidation phase, hydration phase, oxidation phase, and thiolysis. They are little modified in the new approach.

Transfer of Fatty Acids from Fats

Fatty acid (R-CO-OH) is transferred from triglycerides from close protein spirals probably by K vitamins of close protein spirals to empty CoA of close protein spirals. Their CoA coenzymes attract fatty acids for their acyl groups and transform them into R-CO-S-CoA.

Oxidation phase

The beta-oxidation starts in the alpha and the beta carbons of fatty chains that are the carbons closest to COOH.

CH3-CH2 -CH2-CH2-CH2-CH2-CH2-CH2-CH2-CH2-CH2-CH2-C - C - C - C-OH

(ω)   γ β α

The carbon close to the COOH group is marked α, β, or γ.

The distant carbon is marked ω.

NAD coenzymes of close protein spirals attract and oxidize hydrogen atoms at the alpha and beta carbons of fatty acids during polarization and transform into NAD-2H+. The NAD coenzymes transform fatty acids into precursor R-CH2-HC = CH-CO-S-CoA. NAD-2H+ coenzymes transfer hydrogen ions to FAD coenzymes that transform into FAD-2H+ also during polarization.

Hydration Phase

Water hydrate double-bonds with the above molecules and transforms them into fatty chains with hydroxyl group R-CH2-HCOH-CH2-CO-S-CoA.

Oxidation Phase

NAD coenzymes from close protein spirals attract and oxidize hydrogen atoms of the alpha and beta carbons of the above fatty acids with hydroxyl group R-CH2-HCOH-CH2-CO-S-CoA and transform them into R-CH2-CO-CH2-CO-S-CoA also during polarization. The NAD-2H+ coenzymes transfer hydrogen to FAD coenzymes that transform into FAD-2H+ also during the polarization.

Thiolysis Phase

R-CH2-CO-CH2-CO-CoA->R-CH2-CO-CoA+CH3-CO-CoA

The bonds between alpha and the beta carbons of R-CH2-CO-CH2-CO-S-CoA undergo thiolysis (breaking of acetates from the rest of the fatty acids) according to the literature without any logical explanation as to how this happens. The rationale is that folic acid coenzymes attract fatty acids for their methyl groups and that CoA coenzymes attract them for the carboxyl group. They break them into new fatty acids attracted by folic acid (folic acid R-CH2-COO-) and acetates attracted by CoA (CH3-CO-CoA) without thiolysis.

Fatty acids attracted at opposite ends with folic acid and CoA break the weakest bonds between alpha and beta carbons of the fatty acids due to movement of the protein spirals. Folic acid coenzymes transfer the new fatty acids to empty CoA coenzymes of close protein spirals for further decomposition according to the new approach. The same cycle of beta-oxidation of new fatty acids also repeats among close protein spirals during polarization.

Fats and fatty acids exist more inside polypeptide spirals despite the fact that coenzymes with metabolites move around polypeptides because they are very long and heavy molecules. A prove for this approach is the sandwich structure of cell membrane which has outside proteins and inside fats. Polypeptide spirals with fats and fatty acids build cell membranes as parts of attracted and pressed filaments by the outside environment. This approach will be clear after explanation of the structures of prokaryotic and eukaryotic cells.

Chapter 6

Coenzymes of Polypeptides and Amino Acids and Porphyrins

B$_6$ and the other coenzymes are involved in the metabolism of amino acids; they synthesize or decompose amino acids by moving protein spirals in the strings. B$_6$ coenzymes metabolize and transfer amino acids to tRNA and to different coenzymes (B$_1$, folic acid, CoA, and biotin). They transfer also amino groups to C2 of glucose molecules and transform them into NAM and NAG together with specific metabolites of other coenzymes. They transfer glycine to succinyl-CoA that transforms into delta-aminolavulinat. (See the Krebs cycle.) Delta-aminolavulinats are attracted by biotin coenzymes of the second and sixth polypeptides during polarization. The two delta-aminolavulinats fuse into one porphobilinogen also during polarization.

5 - Aminolavulinat   Porphobilinogen

Carboxyl groups (-COOH) of porphobilingen molecules are also attracted by biotins of specific polypeptides. See the picture on the next page.

The four porphobilinogens molecules in biotins of close protein spirals fuse into the porphyrin rings in protein spirals of the strings by Co+++ of B_{12} during depolarization (hypothesis). The porphyrin rings as electronegative complex attract Fe++ ions also released from the unknown molecules (the specific polypeptides or B_{12} or unknown coenzymes) during depolarization.

Porphyrin-Fe++ (Mg++) complexes appear during depolarization, and their function is to oxidize hydrogen ions released from FMN during depolarization. Porphyrin-Fe++ complexes attract oxygen molecules and split them into atoms (O2-> 2O) at aerobic and facultative prokaryotic cells. Oxygen atoms attract hydrogen atoms and transform into water. The porphyrin-Fe++ complexes attract negative ions of the acids from the Krebs cycle that attract hydrogen atoms released from FMN mainly at facultative and anaerobic prokaryotic cells. Why some prokaryotic cells are aerobic, facultative, or anaerobic will be explained after I explain the movements of the protein spirals.

Porphobilinogen   Porphyrin

Porphyn is the simplest porphyrin.[7] Fe++ or Mg++ ions of porphyrin rings are involved in oxidation of H+ ions during depolarization.

Coenzymes and nucleotides are most likely synthesized and decomposed in amino acids of polypeptides that attract them by help of coenzymes that attract metabolites according to the new approach. Their synthesis and decomposition is not discussed in this text because they look for analysis of structures of amino acids and their attractions of different metabolites. The analysis of structure of amino acids and their attractions of different metabolites look for facts from the literature or from the new experiments.

CHAPTER 7
FACULTATIVE, AEROBIC, AND ANAEROBIC PROKARYOTIC CELLS

Porphyrin-Fe++ complexes appear along protein spirals of strings of filaments of all prokaryotic cells during depolarization and disappear during polarization according to the new approach. They enable neutralization of hydrogen atoms, the major waste products of metabolism of different compounds.

Porphyrin-Fe++ complexes are most likely present in biotin coenzymes at different prokaryotic cells during depolarization because the precursors of porphyrin-Fe++ complexes are also made and connected with biotin coenzymes also according to the new approach. The porphyrin-Fe++ complexes attract oxygen molecules and electrically negative parts of acids from the Krebs cycle.

Specific concentrations of the porphyrin-Fe++ complexes along the protein spirals of the strings of filaments of aerobic and facultative prokaryotic cells split oxygen molecules into atoms that neutralize hydrogen atoms according to the new approach. The new cycle of polarization cannot start without neutralization of hydrogen atoms because hydrogen atoms block the attraction power of NAD and FAD coenzymes that attract hydrogen atoms from different metabolites during metabolism.

NAD and FAD coenzymes must be empty so metabolism of prokaryotic cells can take place. The concentrations of the porphyrin-Fe++ complexes along the strings of filaments of anaerobic prokaryotic cells cannot break oxygen molecules. These prokaryotic cells produce enough of the different metabolites from the Krebs cycle. The attracted metabolites from the Krebs cycle as negative ions neutralize hydrogen atoms according to the new approach. After neutralization of hydrogen atoms by oxygen atoms or metabolites from the Krebs cycle, porphyrin-Fe++ complexes disassemble and a new cycle of polarization starts (appendix XIII).

Aerobic and facultative prokaryotic cells have concentrations of porphyrin-Fe++ complexes that break oxygen molecules into atoms and ions according to the new approach.

$$O2 \rightarrow O-+ O-, \ O-+ 2 H+ \rightarrow H2O$$

The porphyrin-Fe++ complexes and attracted oxygen atoms as ions attract H+ ions and neutralize them in water molecules. If H+ atoms are not neutralized, NAD coenzymes cannot take H+ atoms from metabolites and thus a new cycle of polarization cannot does not occur.

Aerobic prokaryotic cells are killed under anaerobic conditions because aerobic prokaryotic cells do not have enough electrically negative metabolites from the Krebs cycle to neutralize H+ ions, which are released also during depolarization.

Anaerobic prokaryotic cells have concentrations of porphyrin-Fe++ complexes that cannot break oxygen molecules into atoms according to the new approach. If anaerobic prokaryotic cells are under anaerobic conditions, their porphyrin-Fe++ complexes attract electrically charged negative metabolites from the Krebs cycle and neutralize them producing the mixed acids. Anaerobic prokaryotic cells produce enough electrically negative metabolites from the Krebs cycle to neutralize H+ ions.

If anaerobic prokaryotic cells are under aerobic conditions, their porphyrin-Fe++ complexes also attract oxygen molecules but without their breaking into atoms. Attracted oxygen molecules prevent attraction of electrically negative metabolites from the Krebs cycle that neutralize H+, and the new cycle of polarization cannot take place. It is a reason they are killed under aerobic conditions.

Facultative prokaryotic cells have concentrations of porphyrin-Fe++ complexes that can attract and split oxygen molecules, so they can live under aerobic conditions. They can live also under anaerobic conditions because they have enough electrically negative metabolites from the Krebs cycle to neutralize hydrogen atoms.

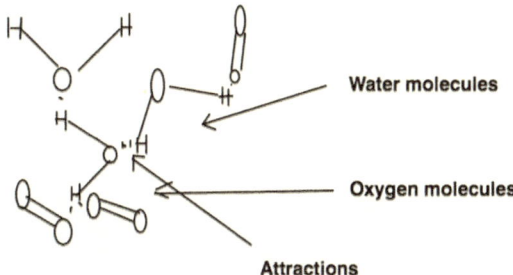

Water molecules

Oxygen molecules

Attractions

Oxygen molecules are dissolved in water. Hydrogen atoms of water molecules attract oxygen atoms of oxygen molecules by electric attractions. Water molecules and oxygen molecules are bipolar.

Chapter 8
Opposite Movements of Protein Spirals

The two protein spirals of the strings move in opposite directions due to the interaction of their polypeptides. The movement causes decomposition and synthesis of metabolites simultaneously. The specific polypeptides of close protein spirals must have a strict order in their attractions and different concentrations of the same coenzymes with their metabolites to move the protein spirals in opposite directions. The strict order of the attraction enables interaction among specific coenzymes and their metabolites, movement of the protein spirals, polarization, depolarization, and metabolism. See the below tables that show interactions among the specific polypeptides.

polypeptide 1	polypeptide 2	polypeptide 3	polypeptide 4	polypeptide 5	polypeptide 6	polypeptide 7	polypeptide 8
attractions of the opposite electrical ions	attractions of the opposite electrical ions	attractions of the opposite electrical ions	attractions of the opposite electrical ions	attractions of the opposite electrical ions	attractions of the opposite electrical ions	attractions of the opposite electrical ions	attractions of the opposite electrical ions
polypeptide 4	polypeptide 3	polypeptide 2	polypeptide 1	polypeptide 8	polypeptide 7	polypeptide 6	polypeptide 5

The first polypeptides with their coenzymes and metabolites of the protein spiral attract the fourth polypeptides with their coenzymes and their metabolites of the other protein spiral and vice versa. The second polypeptides with their coenzymes and metabolites of the protein spiral attract the third polypeptides with their coenzymes and their metabolites of the other protein spiral and vice versa and so on according to the table.

Attractions of the specific polypeptides and new attractions of polypeptides of the protein spirals enable their movement in opposite directions. The logic for this approach is that proteins can move each other only if they have interactions among the specific polypeptides with specific coenzymes that attract specific metabolites.

Opposite movements of protein spirals enable synthesis and decomposition of metabolites simultaneously. One protein spiral decomposes more carbohydrates and synthesizes more fats and amino acids whereas the other synthesizes more carbohydrates and decomposes more fats and amino acids. The protein spiral that decomposes fructose molecules is the decomposition one whereas the other spiral that synthesizes fructose molecules is the synthetic one. The two protein spirals have different polypeptides with different lengths that attract different concentrations of coenzymes and their metabolites; this is necessary for decomposition and synthesis of the same metabolites. If polypeptides have longer lengths, they attract bigger concentrations of coenzymes. Shorter polypeptides attract smaller concentrations of coenzymes.

The picture of two wheels depicts the perpendicular cut of the two close polypeptide spirals that attract each other during metabolism. One of them is the synthetic spiral and the other is the decomposition one. The teeth of the wheels depict the perpendicular cuts of the specific polypeptides that attract each other by metabolites attracted by coenzymes. Polypeptide and protein spirals of strings move each other in opposite directions during attraction and repulsion of metabolites.

Decomposition and biosynthesis of fructose molecules, fatty acids, fats, and amino acids take place among coenzymes according to strict metabolic order; metabolites of different molecules move from coenzyme A to coenzyme B, from coenzyme B to coenzyme C, and so on.

Strict connections among coenzymes, metabolites, and polypeptides of decomposition and synthetic protein spirals are explained by decomposition and synthesis of fructose molecules, NAM acids, NAG, fatty acids, fats, amino acids, and other molecules. The metabolites fructose and fatty acids are also the metabolites of amino acids, coenzymes, and nucleotides. Different quantities of metabolites make different molecules during synthesis and decomposition simultaneously. The strict metabolic order among the specific coenzymes during decomposition of fructose molecules is this:

ATP-> B_1-> folic acid-> biotin.

Metabolites of ATP coenzymes are connected with B_1 coenzymes that attract and take pyruvate metabolites from them. B_1 vitamins with pyruvate metabolites are connected with folic acids that attract acetate molecules from them and transfer them to the metabolites of biotin, CoA, and B_6 coenzymes that attract them also during synthesis of fructose molecules. Acetate metabolites fuse with oxalacetate attracted by biotins and transform them into citrates on biotins. Citrates get in the Krebs cycle in which the different metabolites are produced. The strict metabolic order among the specific coenzymes during biosynthesis of fructose is this:

ATP <-B_1 <-folic acid <-biotin.

Oxalacetate molecules attracted by biotin coenzymes transform into fumarate, succinate, alpha-ketoglutarate, and citrate molecules among biotins by addition of carboxyl groups in the opposite of the Krebs cycle. Citrate metabolites decompose into oxalacetate and acetate molecules between biotins and folic acids. Folic acids attract acetate molecules. B_1 coenzymes attract acetate molecules from folic acids and fuse them with their carboxyl groups into pyruvate molecules. ATP coenzymes attract pyruvate molecules from B_1 and transform them into the intermediary metabolites of fructose molecules. The strict metabolic order among the specific coenzymes during synthesis of fatty acids and fats is this:

folic acid-> CoA-> K vitamins-> ATP.

CoA coenzymes attract acetates from folic acid coenzymes and transform them into fatty acids during decomposition of fructose molecules. K vitamins attract fatty acids from CoA during depolarization and transfer them to 3-phosphoglycerate producing fats during polarization. The strict metabolic order among the specific coenzymes during decomposition of fats and fatty acids is this.

folic acid <-CoA <-K vitamins <-ATP

K vitamins probably attract fatty acids from fats attracted by ATP. CoA attracts fatty acids from K vitamins. Folic acids take acetyl from CoA during decomposition of fatty acids. The strict metabolic order among the specific coenzymes during synthesis of amino acids is this:

ATP, B_1, folic acid, biotin-> B_6.

B_6 with amino groups attracts metabolites from ATP, B_1, folic acid, and biotin and form different amino acids. The strict metabolic order among the specific coenzymes during decomposition of amino acids is this:

ATP, B_1, folic acid, biotin <-B_6.

ATP, B_1, folic acid, and biotin coenzymes attract metabolites from different amino acids attracted by B_6 coenzymes and leave amino groups to B_6.

The strict metabolic orders among the specific coenzymes and the metabolites during decomposition and synthesis of fructose, fatty acids, fats, amino acids, and the other molecules are summarized at bellow tables.

Decomposition of fructose molecules in the decomposition protein spirals
ATP-> B_1-> folic acid-> biotin
folic acid <-CoA <-K <-ATP
ATP, B_1, folic acid, biotin <-B_6

Decomposition of fructose molecules is connected with biosynthesis of fatty acids, fats, and amino acids because decomposition metabolites of fructose molecules are essential for biosynthesis for fatty acids, fats, and amino acids.

Biosynthesis of fructose molecules in the synthetic protein spirals
ATP <-B_1 <-folic acid <-biotin
folic acid-> CoA->K-> ATP
ATP, B_1, folic acid, biotin-> B_6

Decomposition of fatty acids, fats, and amino acids is connected with biosynthesis of fructose molecules because decomposition metabolites of fats, fatty acids, and amino acids are essential metabolites for the synthesis of fructose molecules.

Transformation of many specific metabolites among coenzymes takes places by NAD and FMN coenzymes during polarization. Energy is released during decomposition of fructose, fatty acids, and other complex molecules into simple ones. ATP coenzymes capture their released energy. ATP coenzymes also give off energy during synthesis of fructose, fatty acids, and other complex molecules, which have more energy than do simple ones because they need more energy to hold more atoms in the molecules than do simple ones. Complex molecules need energy during synthesis from simple ones and release energy during decomposition.

Organisms that capture the Sun's energy can make ATP and complex molecules from simple ones. The process of capturing energy in and releasing energy from ATP will be explained later. Organisms that do not use the Sun's energy must decompose complex molecules more than synthesize complex ones because complex molecules provide energy for their metabolism and life.

The movement of intermediate metabolites among coenzymes at protein spirals takes place according to strict metabolic order during decomposition of bigger molecules into smaller ones or during synthesis of smaller molecules into bigger ones. The metabolites' movements are possible only if forces of attraction exist among the coenzymes of synthetic and decomposition protein spirals.

The attraction forces among specific coenzymes for metabolites during decomposition of fructose molecules and synthesis of fatty acids are opposite the movements of metabolites among coenzymes. They look like this (bold letters present the attraction force).

ATP < B$_1$ < folic acid < CoA or biotin; folic acid < CoA < K < ATP; B$_1$, folic acid, biotin < B$_6$
ATP –> B$_1$ –> folic acid –> CoA or biotin; folic acid-> CoA-> K-> ATP; B$_1$, folic acid, biotin-> B$_6$

- Attraction of B$_1$ for metabolites is greater than that of ATP for metabolites, so pyruvate molecules can move to B$_1$.
- Attraction of folic acid for metabolites is greater than that of B$_1$ for metabolites, so acetate molecules can move to folic acid.
- Attraction of CoA for metabolites is greater than that of folic acid for metabolites, so acetate molecules can move to CoA.
- Attraction of K vitamins for metabolites is greater than that of CoA for metabolites, so fatty acids can move to K vitamins.
- Attraction of ATP for the metabolites is greater than that of K vitamins for metabolites, so fatty acids can move to 3-phosphoglycerate attracted by ATP.

Attraction among coenzymes for metabolites during synthesis of fructose molecules and decomposition of fatty acids are opposite the movement of metabolites among coenzymes. They look like this (the thin letters present the movements of the metabolites).

ATP > B$_1$ > folic acid > CoA or biotin; folic acid > CoA > K >ATP; B$_1$, folic acid, biotin > B$_6$
ATP <-B$_1$ <-folic acid <-CoA or biotin; folic acid <-CoA <-K <-ATP; B$_1$, folic acid, biotin <-B$_6$

- Attraction of folic acid for metabolites is greater than that of CoA or biotin for metabolites, so acetate molecules can move to folic acid.
- Attraction of B$_1$ for metabolites is greater than that of folic acid for metabolites, so acetate molecules can move to B$_1$.

- Attraction of ATP for metabolites is greater than that of B_1 for metabolites, so pyruvate molecules can move to ATP.
- Attraction of K vitamins for specific metabolites is greater than that of ATP for metabolites, so fatty acids can move to K vitamins.
- Attraction of CoA for metabolites is bigger than that of K vitamins for metabolites, so fatty acids can move to CoA.

Attraction of coenzymes for metabolites is connected with the quantity of coenzymes. If coenzymes have more quantity than other coenzymes of the strict metabolic order, they have stronger attraction for metabolites than do other in metabolic connections.

In summary, decomposition and synthesis of metabolites depend on attraction of specific coenzymes that attract them. Decomposition and synthesis of metabolites take place in decomposition and synthetic protein spirals of strings simultaneously.

The Zigzag Attractions among Polypeptides

Polypeptides of the strings' protein spirals have specific relationships of coenzymes and metabolites during movement. Position and quantitative concentration of coenzymes and metabolites in the polypeptides of the protein spirals enable the specific polypeptides to attract each other by opposing electrical charges of their metabolites. Different concentrations of coenzymes and their metabolites enable synthesis and decomposition of NAG, NAM, and different metabolites simultaneously.

polypeptide 1	polypeptide 2	polypeptide 3	polypeptide 4	polypeptide 5	polypeptide 6	polypeptide 7	polypeptide 8
ATP a. (--)	B_1, biotin, CoA c. (+)	B_6, Fol. e. (-)	NAD, FAD g. (+)	ATP i. (--)	B_1, biotin, CoA k. (+)	B_6, Fol. K m. (-)	NAD, FAD o. (+)
NAD, FAD b. (+)	K d. (-)	B_{12} f. (+)	ATP h. (--)	NAD, FAD j. (+)	K, biotin l. (-)	B_{12} n. (+)	ATP p. (--)

Coenzymes as bipolar molecules have electrical charges that attract opposite electric charges of bipolar metabolites	Metabolites as bipolar molecules have electrical charges that attract opposite electric charges of bipolar coenzymes (appendix X)
ATP+	PO4--
FAD-, NAD-	H+
B_1+	pyruvate+
biotin+	Different acids of the Krebs cycle+
CoA	acetyl+
K vitamins	fatty acids-
B_6	amino acids-
folic acid	acetyl-
B_{12}	Fe++

Movement of protein spirals in opposite directions is possible only if polypeptides of synthetic protein spirals interact with polypeptides of decomposition protein spirals and if quantitative relationships exist among metabolically connected coenzymes for decomposition and synthesis of metabolites.

Metabolites of close polypeptide spirals of opposite moving protein spirals jump to each other in opposite directions. This is referred to as the zigzag model because metabolites move in opposite directions along the protein spirals in such a pattern.

The zigzag model is presented in the below schemes of the two opposite models. The first model has these quantitative relationships among the specific coenzymes.

ATP < B_1 < folic acid < biotin, folic acid < CoA < K vitamins < ATP and B_1, folic acid, biotin < B_6.

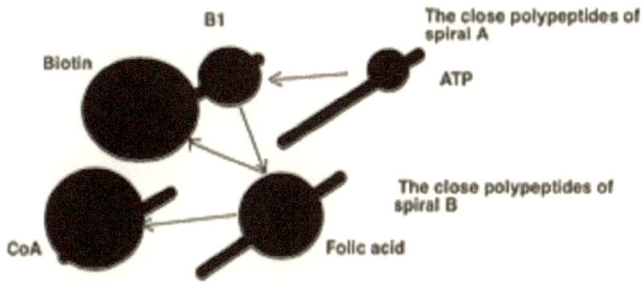

ZIGZAG MOVEMENTS OF METABOLITES AMONGST THE CLOSE POLYPEPTIDES

First model

The first model presents interactions of the specific coenzymes with metabolites of the close polypeptides of decomposition spirals and the close polypeptides of synthetic spirals. Concentrations of specific coenzymes (ATP, B_1, folic acids, and biotins) enable decomposition of polysaccharides and fructose in acetate metabolites that are further transferred in biotins and CoA. Concentrations of specific coenzymes (CoA, K vitamins, and ATP) enable biosynthesis of fatty acids and fats. Concentrations of specific coenzymes (B_6 and the other specific coenzymes) also enable synthesis of amino acids.

The second model has these quantitative relationships among the specific coenzymes.

ATP > B_1 > folic acid > biotin, folic acid > CoA > K vitamins > ATP and B_1, folic acid, biotin < B_6.

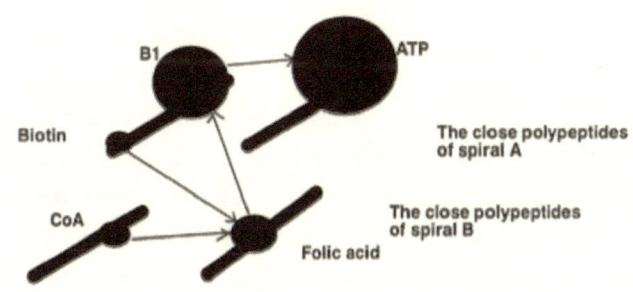

ZIGZAG MOVEMENTS OF METABOLITES AMONGST THE CLOSE POLYPEPTIDES

Second model

The second model also presents interactions of coenzymes with metabolites of the close polypeptides of the synthetic spiral and the close polypeptides of the decomposition spiral.

Concentrations of specific coenzymes (ATP, B_1, folic acids, and biotin) enable synthesis of polysaccharides and fructose from pyruvate molecules produced from acetate molecules and carboxyl groups in B_1 coenzymes. Folic acids attract acetate from citrate molecules in biotin coenzymes. Concentrations of specific coenzymes (folic acids, CoA, and K vitamins) enable decomposition of fatty acids and fats, and concentrations of specific coenzymes (B_6 and other coenzymes) enable decomposition of amino acids.

Protein spirals have the first and second model that enables biosynthesis and decomposition of metabolites simultaneously. Strings can have a prevailing synthesis of carbohydrates and decomposition of fats and amino acids or a prevailing decomposition of carbohydrates and synthesis of fats and amino acids.

Protein spirals of strings of filaments of prokaryotic (eukaryotic) cells move each other as a unified system.

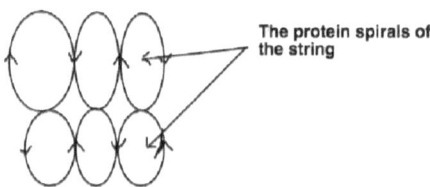

The protein spirals of the string

The protein spirals of the strings of the filaments move each other as a unified system.

Strings of different prokaryotic and eukaryotic cells that have a prevailing of decomposition or synthesis of fructose molecules and synthesis or decomposition of fatty acids, fats, and amino acids of strings of different prokaryotic and eukaryotic cells have the same polypeptides of different concentrations of coenzymes and their metabolites. They also have the same polypeptides organized in groups that work in specific order during decomposition of fructose molecules and synthesis of fatty acids and fats or during synthesis of fructose molecules and decomposition of fatty acids and fats.

Specific polypeptides with coenzymes and metabolites of polypeptide spirals of synthetic and decomposition protein spirals interact in the eighth subgroups of the two specific polypeptides of the interacting polypeptide spirals according to an order—the first, second, third, and so on through the eighth. These are the metabolic groups that repeat along synthetic and decomposition protein spirals during movement of the spirals in opposite directions. They are different because they have different polypeptides with different concentrations of the same coenzymes and metabolites.

The different metabolic groups also have different powers of attraction among their specific polypeptides and consequential order that enable movement of spirals. After the metabolic group A comes the metabolic group B, and then C, and so on.

first subgroup	second subgroup	third subgroup	fourth subgroup	fifth subgroup	sixth subgroup	seventh subgroup	eighth subgroup
polypeptide 1	polypeptide 2	polypeptide 3	polypeptide 4	polypeptide 5	polypeptide 6	polypeptide 7	polypeptide 8
attractions of the opposite electrical ions	attractions of the opposite electrical ions	attractions of the opposite electrical ions	attractions of the opposite electrical ions	attractions of the opposite electrical ions	attractions of the opposite electrical ions	attractions of the opposite electrical ions	attractions of the opposite electrical ions
polypeptide 4	polypeptide 3	polypeptide 2	polypeptide 1	polypeptide 8	polypeptide 7	polypeptide 6	polypeptide 5

Subgroups change according to a consequential order: the first through the eighth, back to the first, and so on.

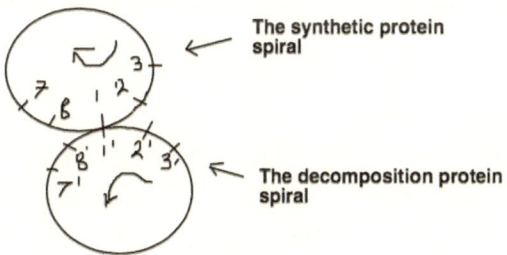

The synthetic protein spiral

The decomposition protein spiral

The order of the eight subgroups causes movement of polypeptide and protein spirals in opposite directions by each other with synthesis of fructose molecules and decomposition of fats, fatty acids, fats, and amino acids or with decomposition of fructose molecules and synthesis of fatty acids, fats, and amino acids.

The subgroups cannot change according to this consequential order: the eighth through the first and back to the eighth, because the power of attraction of the specific polypeptides with coenzymes and metabolites of the subgroups becomes greater from the first subgroup to the eighth.

subgroup 1 <	subgroup 2 <	subgroup 3 <	subgroup 4 <	subgroup 5 <	subgroup 6 <	subgroup 7 <	subgroup 8

Attraction among polypeptides of the eight subgroups become greater because polypeptides of the subgroups have greater concentrations of coenzymes and specific ions that produce greater attraction powers.

Subgroup 1 of the new group comes after attraction of the specific polypeptides of subgroup 8 even though the attraction powers of subgroup 8 of the previous group are greater than the attraction powers of the subgroup 1 of the new group. The rationale for this is that protein spirals move as a unified system that enables attraction of different subgroups simultaneously; this attraction prevails over attractions among subgroups 8 and 1.

The interaction among polypeptides of subgroups means that coenzymes with metabolites of a specific electrical charge attract and take metabolites of coenzymes with metabolites with the opposite electrical charge. The rationale for the attraction powers among the subgroups enables an explanation of biosynthesis of NAM and NAG molecules of peptidoglycan and other complex molecules.

Coenzymes B_1 of polypeptides of the second subgroup after attraction have enough pyruvate metabolites necessary to make NAM and other complex molecules because NAM and the other complex molecules need pyruvate molecules. Coenzymes B_1 of polypeptides of the sixth subgroup after attraction do not have enough pyruvate metabolites necessary for making NAM and other complex molecules, so the attraction produces NAG.

Coenzymes of polypeptides of the subgroups attract and accumulate metabolites during movement of the protein spirals, which is called polarization. When coenzymes of polypeptides cannot attract more metabolites, they release them, which is called depolarization. Polarization and depolarization are discussed in chapter 9.

CHAPTER 9
POLARIZATION, DEPOLARIZATION, AND BIOELECTRICITY

Protein spirals of strings move each other in opposite directions by collecting ions attracted by coenzymes of polypeptides during polarization. Coenzymes of the polypeptides attract ions with specific electrical charges due to their energy.

Polarization occurs when synthetic and decomposition protein spirals attract and accumulate metabolites with electrical charges. Their metabolites with electrical charges spread polypeptides and protein spirals and accumulate their potential energy. Synthesis of polypeptides, nucleic acid pieces, and other metabolites and decomposition occur during polarization.

Depolarization occurs when coenzymes of synthetic and decomposition protein spirals release accumulated metabolites during polarization. They cannot attract them because their coenzymes do not have the power to do so. When the spread protein spirals release metabolites, they collapse and transform their potential energy into kinetic one captured by AMP coenzymes. They collapse due to losing the attraction force of coenzymes of polypeptides due to the accumulation of ions with energy that neutralizes the energy of coenzymes.

Coenzymes of inner polypeptides and their protein spirals attract ions. Opposite electrical charges of coenzymes and metabolites of close polypeptides spirals attract each other making the protein spirals very elastic. Specific coenzymes attracting more specific ions spread out the polypeptide and protein spirals and collect more potential energy that is released during depolarization.

Energy is also collected in decomposition protein spirals by decomposition of complex molecules into simple ones during polarization.

Polarization and depolarization occur in a very short time and move metabolites with electrical charges along protein spirals and produce bioelectricity at prokaryotic cells. The table below shows connections of polypeptides and coenzymes with specific ions-metabolites.

polypeptide 1	polypeptide 2	polypeptide 3	polypeptide 4	polypeptide 5	polypeptide 6	polypeptide 7	polypeptide 8
ATP a. (--)	B_1, biotin, CoA c.(+)	B_6, Fol. e. (-)	NAD, FAD g. (+)	ATP i. (--)	B_1, biotin, CoA k. (+)	B_6, Fol. K m. (-)	NAD, FAD o. (+)
NAD, FAD b. (+)	K d. (-)	B_{12} f. (+)	ATP h. (--)	NAD, FAD j. (+)	K, biotin l. (-)	B_{12} n. (+)	ATP p. (--)

Coenzymes as bipolar molecules have electrical charges that attract opposite electric charges of bipolar metabolites	Metabolites as bipolar molecules have electrical charges that attract opposite electric charges of bipolar coenzymes (appendix X)
ATP+	PO4--
FAD-, NAD-	H+
B_1+	pyruvate+
biotin+	different acids of the Krebs cycle+
CoA	acetyl+
K vitamins	fatty acids-
B_6	amino acids-
folic acid	acetyl-
B_{12}	Fe++

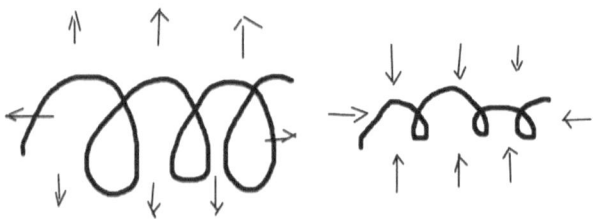

The protein spirals of the strings spread during polarization and they collapse during depolarization.

Coenzymes of polypeptides and protein spirals collect ions until they cannot attract more. They then release them due to lack of energy to hold them. When they do so, they disturb other coenzymes with lack of energy to hold more ions. This release of ions produces a domino effect that collapses the spirals; this is depolarization.

AMP and other coenzymes capture the released kinetic energy, increase movement of their atoms, and change their electrical charge from negative to positive, which attracts negative PO4---and increases kinetic energy. By attracting PO4---groups, AMP transforms into ADP and then into ATP.

Other coenzymes attract the kinetic energy with faster movements of their atoms, which enables them to collect more ions during the new cycle of polarization. Attraction energy of the majority of coenzymes is spent during attraction of ions during polarization. ATP coenzymes help coenzymes maintain their levels of energy by releasing their PO4---groups during polarization. ATP by releasing of PO4---groups release energy that is transferred to kinetic energy to coenzymes via protein spirals. ATP by releasing PO4---groups transform into ADP and AMP. Any form of energy is always connected to matter as they are the same.

After depolarization, coenzymes of the spirals renew their energy, attract ions, and start a new cycle of polarization. Their energy is reduced during polarization due to biosynthetic reactions that spend it. ATP is also produced during polarization where some complex molecules are split and release energy, which is captured by AMP or ADP. ATP is produced more during depolarization than polarization at the cells that do not use the extra energy.

The protein spirals transfer energy from ATP to the coenzymes or from the coenzymes to AMP the same as the wires from the electric producers to houses.

Polarization and depolarization of filaments are synchronized events that repeat alternatively in the strings of filaments and produce bioelectricity phenomena of prokaryotic cells. Depolarization waves start from coils of strings and spread to their straight parts.

Filaments of prokaryotic cells have straight and coiled parts of strings at both sides. They start from the coil parts because metabolic activity is greater there than at the straight ones because coenzymes of protein spirals have more energy because the new protein spirals at the coil parts have greater amounts of coenzymes than the protein spirals at the straight parts do.

New protein spirals attract more coenzymes and metabolites than do protein spirals at the straight parts because polypeptides of the new protein spirals are not used during metabolism as is the case with polypeptides of other protein spirals. The greater amounts of coenzymes attract more energy that in turn attracts more ions and reach saturation faster.

Depolarization waves start from opposite poles of prokaryotic cells and finish at the opposite poles almost at the same time. The waves go in opposite directions because one pole has ends of filaments that have the coils and the straight parts of strings.

Frequency of Depolarization

NAD and FMN coenzymes in the coil parts of the strings most likely start depolarization waves first among other coenzymes because they accumulate hydrogen very quickly because they are involved in almost all metabolic processes. This is not a case with other coenzymes, which accumulate ions up to saturation and release them slower than NAD and FMN coenzymes do.

When NAD and FMN coenzymes release hydrogen ions, they disturb and reduce the attraction force of close polypeptides and coenzymes. These coenzymes release ions under their influence and cause depolarization waves that spread to the other coenzymes along the strings from the coil parts to the straight ones. Frequency of depolarization and polarization depends on concentrations of NAD and FMN and concentrations of the other coenzymes in polypeptides. Prokaryotic cells that have higher concentrations of NAD and FMN coenzymes have a higher frequency of depolarization than those with less because higher concentrations of NAD and FMN attract hydrogen ions at a greater rate. That means they take less time to accumulate hydrogen ions and release them.

They speed up metabolism because they are involved in all biochemical reactions. The higher frequencies of depolarization at prokaryotic cells produce fewer complex molecules (peptidoglycan, lipopolysaccharides, teichoic acids, and others) because other coenzymes do not have enough metabolites to produce them because they release their metabolites quickly.

If prokaryotic cells have lower concentrations of NAD and FMN, they have a lower frequency of depolarization because they can attract fewer hydrogen ions, so it takes them longer to accumulate and then release them. Other coenzymes have enough metabolites to produce complex molecules because they do not release their metabolites quickly.

The next picture shows high-and low-frequency waves. The frequency of depolarization depends on concentrations of NAD and FMN.

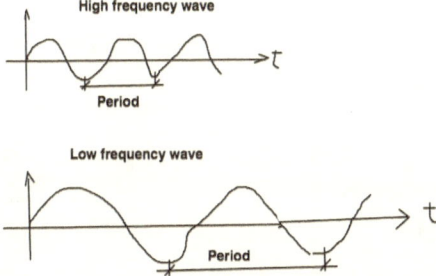

Prokaryotic cells have different frequency of depolarization and polarization that manifest as high or low-frequency waves.

The next picture shows the low-and the high-energy waves of depolarization. The energy of depolarization depends on concentrations of all coenzymes in polypeptides of inside polypeptide spirals of protein spirals of all filaments.

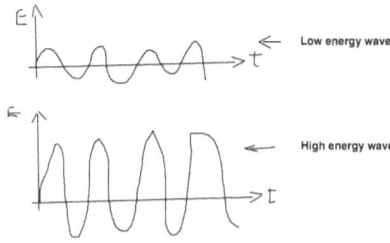

If prokaryotic cells collect fewer ions, they release less kinetic energy. When they collect more ions, they release more kinetic energy that produces high-energy waves.

The frequency waves at prokaryotic cells are the steep line due to the speed of depolarization. Polarization lasts longer than does depolarization.

The frequency waves at prokaryotic cells are shown in the next picture.

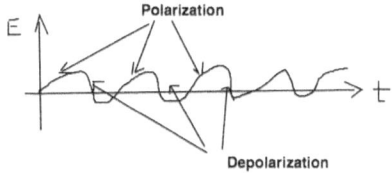

Amplitudes of polarization and depolarization of prokaryotic cells resemble the teeth of a saw. The less-sharp slopes of teeth depict polarization whereas the sharp ones depict depolarization.

Quantitative relationships between NAD and FMN determine the speed and frequency of depolarization. If prokaryotic cells have more NAD than FMN, they metabolize faster and have a higher frequency of depolarization because NAD coenzymes transfer H+ ions to FMN. The opposite is true in terms of less NAD than FNM. Lower concentrations of NAD enable slower accumulation of FMN and more time to trigger depolarization.

Growth of filaments and polypeptide and nucleic acid synthesis can be explained only after the explanation of the new structures, metabolism of the different metabolites, polarization, depolarization, bioelectricity, and other phenomena. They cannot be explained without them because they are explained by them.

CHAPTER 10
GROWTH OF STRINGS AND FILAMENTS

Prokaryotic cells consist of filaments whereas eukaryotic cells consist of complex cylinders, which will be explained later. Filaments have two strings whereas complex cylinders have four. Biosynthesis of protein and nucleic acid spirals of prokaryotic and eukaryotic cells occurs in the strings. Strings interchange parts of protein and nucleic acid spirals inside and outside the cells among pairs of filaments or cylinders. Biosynthesis and interchange of protein and nucleic acid spirals enable growth of the strings, filaments, cylinders, and cells.

Strings and cylinders consist of protein spirals organized as is ds DNA and can attract and hold helicoid strands of ds DNA. The protein spiral of the string that attracts and holds the negative strand of ds DNA by its inner polypeptide spiral is called the negative protein spiral.

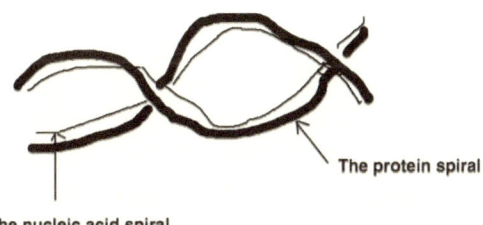

The protein spiral

The nucleic acid spiral

Protein spirals and strands of nucleic acids have the helicoid model.

Protein spiral that attract and hold the positive strand of ds DNA by inner polypeptide spirals is called the positive protein spiral.

Negative and positive protein spirals move each other in opposite directions, synthesize and decompose metabolites, and enable movement of metabolites from biotin to ATP and from ATP to biotin simultaneously. (See chapter 9 for more details.)

Protein spirals that attract more fructose molecules than their simple metabolites are called synthetic protein spirals. The other protein spiral, which attracts more simple metabolites of fructose molecules than fructose molecules, is called the decomposition one. Positive and negative protein spirals can be synthetic or decomposition. If positive proteins spiral are synthetic, negative protein spirals are decomposition and vice versa. This will be explained.

Decomposition or synthetic protein spirals attract genes in positive strands of DNA by polypeptides because they make them during protein and nucleic acid biosynthesis. Filaments have synthetic and decomposition strings although that both strings have synthetic and decomposition protein spirals. Synthetic strings synthesize synthetic protein spirals whereas decomposition strings synthesize decomposition protein spirals. Filaments' two strings have coiled and straight parts on opposite ends.

Filaments have two strings. Each side of filament has straight and coil part of the strings.

Synthetic and decomposition strings have synthetic and decomposition protein spirals. Synthetic strings have decomposition protein spirals as positive protein spiral by which polypeptides attract complementary decomposition genes of positive strands of ds DNA. Complementary genes here mean genes that produce polypeptides and attract each other.

Decomposition strings have synthetic protein spirals as positive protein spirals by which polypeptides attract complementary synthetic genes of positive strands of ds DNA. This will be explained with more details during the explanation of synthesis of decomposition protein spirals in decomposition strings and synthesis of synthetic protein spirals in synthetic strings.

The strings of the filaments or the cylinders do not interchange parts of protein spirals among themselves because they do not have the same parts of protein and nucleic acid spirals. They interchange such parts with other strings of the other filaments or the other cylinders with the same parts of protein and nucleic acid spirals. The two filaments that interchange parts of protein and nucleic acid spirals are called a pair of filaments. The two cylinders that interchange parts of protein and nucleic acid spirals are called a pair of cylinders. Why this is will be made clear after explanation of the synthesis of synthetic and decomposition protein spirals in synthetic and decomposition strings.

The parts of strings that grow and where biosynthesis of synthetic or decomposition protein spirals occur are called coils while the opposite part, which does not grow and where biosynthesis of protein spirals does not take place, are called straight ones. Their functions will be explained in more detail later after some explanation.

Filaments and cylinders undergo growth and division and they enable divisions of the cells. Synthesis and decomposition of metabolites take place during metabolism simultaneously at different prokaryotic and eukaryotic cells. Strings with protein and nucleic acid spirals must exist in prokaryotic and eukaryotic cells that enable synthesis and decomposition of metabolites simultaneously. The strings must also exist with synthetic and decomposition protein spirals because they are necessary for synthesis and decomposition of metabolites. If strings synthesize synthetic protein spirals, the other filament or cylinders must synthesize decomposition protein spirals because synthetic and decomposition protein spirals must be produced for the strings to exist.

String produce either synthetic or decomposition protein spirals but not both due to the new approach to protein synthesis explained in the next chapter. The same number of strings that produce synthetic and decomposition protein spirals must exist to produce strings that grow by interchange of parts of protein and nucleic acid spirals.

Prokaryotic cells grow and undergo binary fission simultaneously, so the strings must be organized in filaments to grow at both ends and divide by the structures of filaments. The logic of the string and filament models will be clear after explanations of the other cell's structures and their phenomena because they are also built by them.

The model of the pair of filaments is created to enable interchange of protein and nucleic acid spiral parts and growth of filaments. All phenomena in nature have causes and effects as is the case with the above models and are due to different qualities and quantities of matter. They cannot exist per se because an empty space without matter does not exist. The different phenomena of the cells depend on each other as a united system.

Interchange of Parts of Protein and Nucleic Acid Spirals

Different filaments of pairs of strings have specific positions of coils and straight parts at their ends due to the interchange of the protein spiral parts. Strings cannot have protein and nucleic acid synthesis and cannot grow without interchanging parts of protein and nucleic acids spirals.

Filament 1

Coil part of the synthetic string	Straight part of the synthetic string
Straight part of the decomposition string	Coil part of the decomposition string

Filament 2

Coil part of the decomposition string	Straight part of the decomposition string
Straight part of the synthetic string	Coil part of the synthetic string

Filament 1 has the coil part of the synthetic string and the straight part of the decomposition string on the left side and the straight part of the synthetic string and the coil part of the decomposition string on the right side; filament 2 has the opposite. The straight parts of synthetic and the straight parts of decomposition strings of filaments must be on the same side due to the interchange of parts of the specific protein spirals. Coil parts of synthetic and decomposition strings do not interchange parts of protein spirals as do their straight parts due to their growths.

Filaments 1 and 2 can interchange parts of protein and nucleic acid spirals. They grow and divide only by the described model. Filaments must have two strings of specific organization to divide. To grow in one direction and have protein and nucleic acid synthesis, the strings must have specific organization of protein spirals that move in opposite directions. Ds DNA is responsible for biosynthesis of synthetic and decomposition protein spirals also according to the new approach.

Before an explanation of biosynthesis of the protein and the nucleic acid spirals, I will explain how specific protein and nucleic acids spiral parts interchange. This interchange does not contradict protein and nucleic acid biosynthesis because they explain them and vice versa.

Synthetic string of filament 1

Synthetic protein spiral >>>	Synthetic protein spiral >>>
<<<*New synthetic protein spiral*	<<< Decomposition protein spiral
Coil part	Straight part

Decomposition string of filament 1

Decomposition protein spiral <<<	Decomposition protein spiral <<<
Synthetic protein spiral >>>	*New decomposition protein spiral* >>>
Straight part	Coil part

Synthetic string of filament 2

Decomposition protein spiral >>>	Decomposition protein spiral >>>
<<< *New decomposition protein spiral*	<<< Synthetic protein spiral
Coil part	Straight part

Decomposition string of filament 2

Synthetic protein spiral <<<	Synthetic protein spiral <<<
Decomposition protein spiral >>>	New synthetic protein spiral >>>
Straight part	Coil part

Synthetic strings of filament 1 synthesize new synthetic protein spirals at the coil parts. Synthetic strings are called synthetic because they synthesize synthetic protein spirals. New synthetic protein spirals at the coil parts stay there and change movement to have the same movement as the synthetic strings. They move along the same string from the coil to the straight part.

Parts of synthetic protein spirals of filament 1 are expelled during depolarization and are attracted by straight parts of decomposition strings of filament 2. Straight parts of synthetic strings of filament 1 and straight parts of decomposition strings of filament 2 are on the same side due to this interchange. They move from the straight parts to the coil parts of decomposition strings and enable polypeptide and nucleic acid biosynthesis and growth of the decomposition strings of filament 2.

Decomposition strings of filament 1 synthesize new decomposition protein spirals at the coil parts. Decomposition strings are called decomposition one because they synthesize decomposition protein spirals. New decomposition protein spirals at the coil parts stay there and change movement to have the same movement as the decomposition strings. They move along the same string from the coil to the straight part.

Parts of decomposition protein spirals of filament 1 are expelled during depolarization and are attracted by straight parts of synthetic strings of filament 2. Straight parts of decomposition strings of filament 1 and straight parts of synthetic strings of filament 2 are on the same side due to this interchange. They move from straight parts to coil parts of the decomposition string and enable polypeptide and nucleic acid biosynthesis and growth of the synthetic string of filament 2. The similar scenario is also for the strings of filament 2.

Synthetic strings of filament 2 synthesize new synthetic protein spirals at the coil parts. Synthetic strings are called synthetic due to their synthesizing synthetic protein spirals. New synthetic protein spirals at the coil parts stay there and change movement to have the same movement as synthetic strings. They move along the same string from the coil parts to the straight parts.

Decomposition strings of filament 2 synthesize new decomposition protein spirals at the coil parts. Decomposition strings are called decomposition because they synthesize decomposition protein spirals. New decomposition protein spirals at the coil parts stay there and change movement to have the same movement as decomposition strings. They move along the same string from the coil to the straight part.

Parts of decomposition protein spirals of filament 2 are expelled during depolarization and are attracted by the straight parts of synthetic string of filament 1. Straight parts of decomposition strings of filament 2 and straight parts of synthetic strings of filament 1 are on the same side due to this interchange. They move from straight parts to coil parts of the decomposition strings and enable polypeptide and nucleic acid biosynthesis and growth of synthetic strings of filament 1.

In summary, the straight parts of the synthetic strings of filament 1 and the decomposition strings of filament 2 of the pair are at the same pole of a prokaryotic cell. Also, the straight parts of decomposition strings of filament 1 and synthetic strings of filament 2 of the same pair are at the same opposite poles of the same prokaryotic cells. The specific positions of the straight parts of the specific strings of filaments of the pair are due to the interchange of the parts of specific protein and nucleic acid spirals. Coil parts of synthetic and decomposition strings of filaments 1 and 2 do not interchange parts of protein and nucleic acid spirals. The different pairs of filaments must have this organization in the different cells due to their growth.

These hypothetical phenomena can exist only if the strings have protein spirals that move in opposite directions by spiral movements. These hypothetical phenomena do not contradict the opposite movement of decomposition and synthetic protein spirals and the growth of filaments at the both opposite directions.

Filaments or cylinders can grow and divide only in the pairs

- if they have two different strings of the specific organization in the pairs described, (Strings grow on coil parts, not straight parts.)
- if strings of the filaments interchange parts of proteins and nucleic acid spirals among the straight parts of the specific strings, (Parts of protein and nucleic acids are expelled from the strings and attracted by protein spirals of the other strings.)
- if the coil parts of the strings do not interchange parts of the protein and nucleic acids spirals due to growths of the coil parts of filaments. (New parts of protein and nucleic acids spirals produced during biosynthesis of protein and nucleic acid spirals spread their diameters at the coil parts where they stay and change movement in the opposite directions. The strings grow because parts of protein and nucleic acid spirals get in at the straight parts of the strings.)

Parts of new protein and nucleic acids spirals are produced at the coil parts of synthetic and decomposition strings and enable growth of strings of different filaments. Coil parts of the strings do not interchange parts of protein and nucleic acids spirals though they produce them. The parts of new protein and nucleic acids spirals are spread and attracted by close protein spirals of the same coil parts. They transform into new protein and nucleic acid spirals at the coil parts and move along the strings by spiral movement. New protein and nucleic acids transform into new protein spirals by attracting metabolites during spiral movement along the strings.

The opposite movement of two protein spirals at specific strings cause overgrowth of specific protein spirals at the straight parts. Decomposition strings have overgrowth of decomposition protein spirals at straight parts and overgrowth of synthetic protein spirals at coil parts. Overgrowth is the product of opposite movement of protein spirals that move each other in opposite directions.

The decomposition string

	Decomposition protein spiral >>>	>>> Overgrowth of decomposition protein spiral
Overgrowth of synthetic protein spiral <<<	<<< Synthetic protein spiral	

Coil part   Straight part

Protein spirals of decomposition strings attract three nucleic acids.

- **Old ds DNA.** ATPs of the polypeptides of the decomposition protein spiral attract negative strands of old ds DNA. Polypeptides of synthetic protein spirals attract positive strands of the old ds DNA. Old ds DNA has a helicoid shape; it is split into positive and negative strands at the coil parts. Other details of old ds DNA are explained more during the discussion of biosynthesis of nucleic acids and polypeptides.
- **New ds DNA.** Negative strands of ds DNA transform into new ds DNA by biosynthesis of new strands of nucleic acids. They move from straight to coil parts by movement of the polypeptides in spirals. They do not have a helicoid shape because decomposition protein spirals attract them only for negative strands via their ATP.

Positive strands of old ds DNA produce new polypeptides that produce tRNA from them. New polypeptides produce new polypeptides and polypeptide spirals and positive nucleic acid pieces that attract each other into bigger pieces that produce complementary nucleic acids and new ds DNA. New ds DNA produced by polypeptides moves from coil to straight parts by movement of polypeptide spirals in the same direction. This ds DNA does not have a helicoid shape because polypeptides of new decomposition protein spirals attract only positive strands. Other details of new ds DNA are explained more during the discussion of biosynthesis of polypeptides and nucleic acids.

- **tRNA with amino acids.** tRNA with amino acids moves with polypeptides of polypeptide spirals from coil to straight parts. They get in and out of strings due to spiral movements of the polypeptide spirals. Other details of tRNA will be explained more during the discussion of nucleic acids and polypeptides as will the winding and unwinding phenomena.

Overgrowth of decomposition protein spirals occurs at straight parts. Overgrowth does not have tRNA with amino acids because they are utilized during biosynthesis of polypeptides. Overgrowth breaks the parts of decomposition protein and nucleic acid spirals, which are expelled from the strings during depolarization. Overgrowth of synthetic protein spirals of decomposition strings at coil parts bring new parts of decomposition protein and nucleic acids spirals, which are spread during depolarization. They stay in the coil parts and move to the straight parts.

Synthetic strings have overgrowth of synthetic protein spirals at straight parts and overgrowth of decomposition protein spirals at coil parts. Overgrowth is a product of protein spirals that move each other in opposite directions.

The synthetic strings

	Synthetic protein spirals >>>	>>> Overgrowth of synthetic protein spirals
Overgrowth of decomposition protein spirals <<<	<<< Decomposition of protein spirals	
Coil part		Straight part

Overgrowths at the straight parts of synthetic or decomposition protein spirals separate from the strings during depolarization whereas overgrowths at coil parts of synthetic or decomposition protein spirals do not separate during depolarization due to the described model of the string.

Decomposition and synthetic protein spirals have characteristics that enable separation of overgrowth at straight parts and prevent separation of overgrowth at coil parts according to the new view. Overgrowth at coil parts of decomposition or synthetic strings have old polypeptide spirals rich with metabolites that do not allow their breaking during depolarization due to strong attraction among polypeptides. New polypeptide spirals and new ds DNA also do not allow breaking of overgrowth at coil parts because they attract each other and make them stronger. Overgrowth at straight parts of decomposition or synthetic strings breaks easily during depolarization because polypeptides of protein spirals do not have enough metabolites to strengthen attraction among the polypeptides of the protein spirals and do not have support as overgrowth at the coil parts does.

Separation of overgrowth of protein spirals at straight parts occurs at the point of weakest attraction among the polypeptides with smaller concentrations of metabolites with their smaller electrical charges, which make them attract each other very weakly. Separated polypeptides at overgrowth attract each other as a lock-and-key mechanism by their metabolites with electrical charges.

The last specific polypeptides of straight parts of protein spirals that move to coil parts attract polypeptides of parts of protein and the nucleic acid spirals by attraction during polarization. Attractions among polypeptides of strings work also as a lock-and-key mechanism that enables the interchange of parts of protein spirals and nucleic acids among strings of the pairs of filaments and their growths. Breaking of overgrowth and interchange of parts of protein and nucleic acid spirals hold straight parts of strings without growth, which is necessary for the logic of the model.

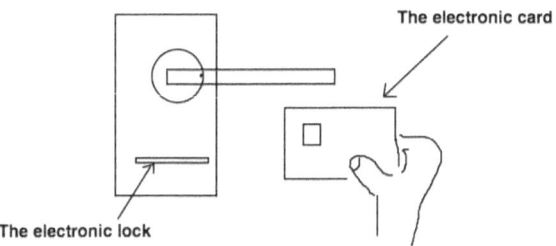

The electronic card and the electronic lock resemble the lock-and-key mechanism that enables attraction among polypeptides of protein spirals.

Some filaments such as pili and flagella interchange parts of protein spirals and nucleic acids outside prokaryotic cells among filaments that belong to one or another prokaryotic cell that belongs to the same species. They easily lose parts of protein spirals and nucleic acids during the interchange; those parts are not attracted and are used by the strings to make biofilms.

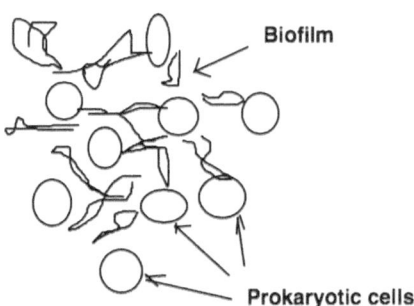

Many prokaryotic cells produce a biofilm that consists of protein spirals, polysaccharides, minerals, and other molecules. Biofilm is a product of the interchange of parts of protein spirals and nucleic acids among prokaryotic cells of the same species. Biofilm provides them water, food, and protection.

Poor interchange of parts of spirals and nucleic acids at some pairs of filaments cause these filaments to not grow; they transform into plasmids, small, circular, double-stranded DNA according to the literature (see appendix XIV). Plasmids consist of two small pieces of ds DNA in two short strings of filament according to the new view. When the cell is killed, the two short filaments with ds DNA attract each other at their ends into circular structures of plasmids according to the new explanation.

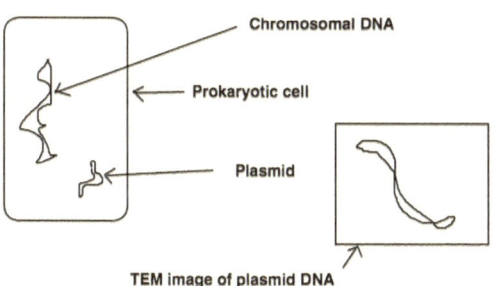

TEM image of plasmid DNA

Circular and small ds DNA are seen at killed prokaryotic cells by TEM.

Jumping genes or transposons exist in prokaryotic cells; they are pieces of DNA that move in genomes of prokaryotic cells (according to the literature), which exist in the strings of filaments and cannot exist without them because they enable the jumping phenomenon. The existence of jumping genes with proteins proves the interchange of parts of protein spirals with pieces of DNA among filaments of prokaryotic cells (appendix XIV). Attraction and movement of ds DNA for protein spirals of strings will be explained in following chapters before the explanation of the organization of the genes and biosynthesis of protein and the nucleic acid spirals.

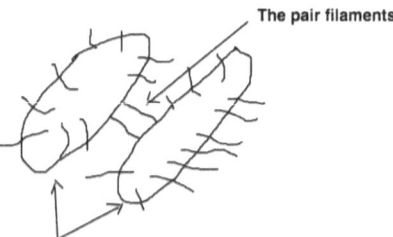

The pair filaments

The same prokaryotic cells

Conjugation occurs among prokaryotic cells of the same species. The pair of filaments between two prokaryotic cells of the same species attracts each other and interchange parts of protein spirals and nucleic acids. Conjugation is also a proof of the interchange of the parts of the protein and nucleic acids spirals.

CHAPTER 11
NEW APPROACH TO BIOSYNTHESIS OF PROTEIN AND NUCLEIC ACID SPIRALS

Movement of protein spirals in opposite directions causes movement of nucleic acids, metabolism of metabolites, biosynthesis of protein and nucleic acid spirals, and string and filament growth. These phenomena are inseparable because they depend on each other.

Protein spirals of strings of filaments have helicoid structures by which they attract and hold the helicoid structures of ds DNA. It moves from the straight to the coil part of synthetic or decomposition strings despite opposite movement of protein spirals that attract their strands in the strings. This explains movement of ds DNA from straight to coil parts.

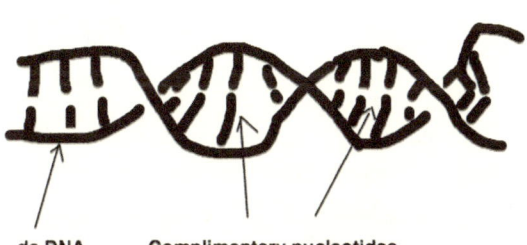

ds DNA   Complimentary nucleotides

Protein spirals of the strings of filaments also have helicoid (rope) structures to attract and hold the helicoid structure of ds DNA.

The ds DNA of strings has positive and negative strands at the coil parts. Positive and negative strands have helicoid shapes due to their attraction to protein spirals in helicoid structures. Positive strands bring genes that are directly translated into polypeptides with involvement of mRNA according to the new approach and new logic. The central dogma does not explain biosynthesis of protein and nucleic acid spirals logically because it does not have any logical connection to the different phenomena of prokaryotic and eukaryotic cells. The new approach negates the central dogma of the protein biosynthesis (appendix XV). The new approach explains biosynthesis of protein and nucleic acids by new structures and connections with different phenomena of prokaryotic and eukaryotic cells in causative relationships.

Before division (binary fission, mitosis), prokaryotic and eukaryotic cells must produce double the amount of nucleic acids so daughter cells have approximately the same amount of nucleic acids as their mother cells immediately after division. After splitting off ds DNA, negative strands synthesize positive strands and new ds DNA, but they make only half the necessary nucleic acids whereas positive strands after splitting off ds DNA make protein spirals and half of nucleic acids before divisions of the cells according to the new approach.

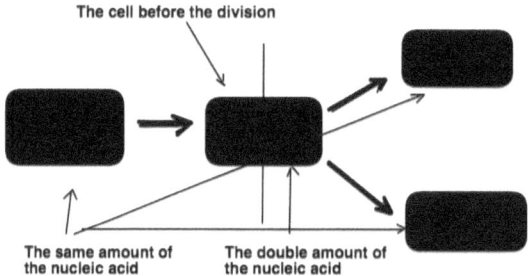

The cells after division have the same amount of nucleic acids. The cells immediately before division must have double the amount of the nucleic acids than the cells after the divisions so the cells after division have the same amount of nucleic acids.

This hypothesis is explained by the facts that daughter cells have amounts of nucleic acids and proteins similar to those their mother cells had during a similar span of life. If not, they should have different bodies and characteristics. This possibility is excluded. The mother cells must produce double the amount of nucleic acids and proteins before the division to satisfy this fact. That means both strands of ds DNA must produce double amounts of nucleic acids the mother cells have before division. The double amounts can be produced only in positive and negative strands of the split ds DNA.

Protein spirals move in opposite directions as a unified system and can biosynthesize protein and nucleic acid spirals only as a unified system.

The coils of the polypeptide spiral of another protein spiral

The coils of the polypeptide spiral of one protein spiral

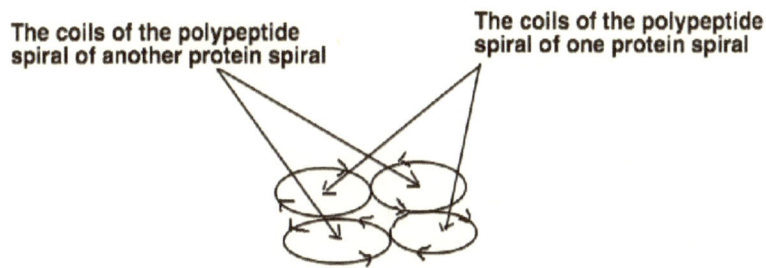

The coils of the polypeptide spirals of the protein spirals move into the opposite directions

Protein spirals attract ds DNA and vice versa. Attraction between ATP of protein spirals and strands lasts for a very short time due to the circular spiral movements of polypeptide spirals in protein spirals. ATP connections spread ds DNA between protein spirals and do not move ds DNA to coils or straight parts. Movement of ds DNA to the coils enables protein and nucleic acid biosynthesis.

Attractions between positive strands of ds DNA and polypeptides of protein spirals allow movement of ds DNA to coil parts according to the new approach. Positive strands of ds DNA attract polypeptides because positive strands have triplets of nucleotides that attract amino acids of polypeptides of protein spirals. Negative strands of ds DNA do not attract polypeptides because they do not have such triplets.

Nucleotides attract each other because they are complementary and close, but they are not permanent attractions because nucleotides of both strands rotate around the axes of strands despite their attractions. That enables attraction of the triplets for amino acids of polypeptides of protein spirals; that attracts triplets of positive strands by specific amino acids and moves the ds DNA to the coil parts.

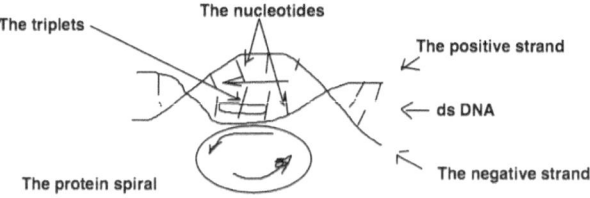

The ds DNA moves in the direction of the circular spiral movements
of the specific polypeptides of the protein spirals

Connections between positive strands of nucleic acids and polypeptides of polypeptide spirals last also for short time due to circular-spiral movements of polypeptides in polypeptide spirals and polypeptide spirals in protein spirals. Connections between positive strands of nucleic acids and polypeptides of polypeptide spirals move nucleic acids spirals in a direction because power does not exist to push the ds DNA in an opposite direction.

Ds DNAs of the helicoid structures is split into negative and positive strands at coils without the unwinding dogma described in the literature (appendix XVI). Ds DNA is split in two strands due to the different sizes of coils of opposite moving protein spirals at coil parts and attractions of strands of ds DNA at the protein spirals. Different sizes are the consequence of different concentrations of metabolites at their polypeptides that cause spreading of their coils. New protein spirals have smaller protein coils than do the coils of old protein spirals. Two split strands of old ds DNA are later separated from old ds DNA after forming new ds DNA because amino acids of new protein spirals that move from coil parts of strings to straight ones attract positive strands of new ds DNA and separate them from old ds DNA.

New protein spirals move from coil parts to straight ones collecting more coenzymes and their metabolites that spread their coils more despite depolarization; depolarization does not remove all their metabolites. Protein spirals move from straight to coil parts after interchanging protein spirals, which collect coenzymes and metabolites that spread their coils more and more despite depolarization. Helicase enzymes enable the splitting of ds DNA into two strands without any logic and connection with protein structures according to the literature.

Positive strands of old ds DNA transform directly into mRNA due to their proximity to protein spirals and their enzymes that transform them, which is a result of attraction among their nucleotides and amino acids of polypeptides. This proximity produces free and attracted mRNA genes simultaneously due to rotation of the nucleotides around their axes.

Free genes of mRNA spirals attract amino acids with their tRNA and make polypeptides according to the new approach. Biosynthesis of protein and nucleic acid spirals according to the new approach is explained in brief below.

The mRNA attracts amino acids with their tRNA triplets and synthesizes newly synthesized polypeptides that attract coenzymes. Newly synthesized polypeptides break tRNA on the attracted triplets and the pieces by the coenzymes.

The newly synthesized polypeptides join attracted triplets in pieces of new ss RNA that attract free nucleotides and synthesize the complementary chain of new ds RNA; these are known as Okazaki fragments. Newly synthesized polypeptides with coenzymes cut the genes of mRNA that produced them and fuse their new ds RNA pieces with pieces of the cut mRNA in incomplete strands of ds RNA. Incomplete strands of ds RNA by the new ds RNA (Okazaki fragments) make complete ds RNA.

Newly synthesized polypeptides and coenzymes with ions attract each other in protein spirals that attract complete ds RNA, which transforms into new ds DNA during depolarization. New protein spirals with new tRNA and ds DNA spirals spread at the coil parts and fuse with protein spirals and the ds DNA and move from coil to straight parts.

The schemes below show two filaments with strings that interchange parts of protein and nucleic acid spirals. The filaments have synthetic and decomposition strings. Synthetic strings produce decomposition protein spirals, parts of decomposition protein spirals, and ds DNA, whereas decomposition strings produce synthetic protein spirals, parts of ds DNA. Synthetic protein strings interchange parts of synthetic spirals and ds DNA with decomposition strings, which interchange parts of decomposition spirals and ds DNA with synthetic strings.

The model of filament has the two strings. Each side of filament has the straight and the coil part of the strings.

Synthetic string of filament 1

>>> Synthetic protein spiral with ds DNA, tRNA, ss DNA of ds DNA >>>

<<< (New synthetic protein spiral with ds DNA and tRNA) + Decomposition protein spiral with ss RNA or ss DNA of ds DNA <<<

Coil part   Straight part

Decomposition string of filament 1

<<< Decomposition protein spiral with ds DNA, tRNA, ss DNA of ds DNA <<<

>>> Synthetic protein spiral with ss RNA + (New decomposition protein spiral with ss RNA or ssDNA of ds DNA) >>>

Straight part   Coil part

Decomposition protein string of filament 2

<<< Decomposition protein spiral with ds DNA and t RNA <<<

>>> Synthetic protein spiral with ss RNA or ss DNA of dsDNA + (New decomposition protein spiral with ds DNA and tRNA) >>>

Straight part   Coil part

Synthetic string of filament 2

>>> Synthetic protein spiral with dsDNA and tRNA >>>

<<< (New synthetic protein spiral with ds DNA and tRNA) +
Decomposition protein spiral with ss RNA or ss DNA of ds DNA
<<<

Coil part   Straight part

New protein spirals with new ds DNA and tRNA move from coil to straight parts and get out from the strings as overgrowth. They spend tRNA with nucleic acids during biosynthesis of protein and nucleic acids. Overgrowth breaks into parts of protein spirals with new ds DNA spirals without tRNA. Strings of the other pair of filaments attract parts of new protein spirals and new ds DNA from strings of the paired filaments.

Protein spirals of strings that move from straight to coil parts attract parts of new protein spirals and new ds DNA spirals. When new ds DNA of spiral structures gets into the strings, opposite-moving protein spirals attract and transform them into helicoid structures (the rope model).

Parts of proteins and new ds DNA enable growths of strings with the described phenomena of doubling the protein and nucleic acid spirals before division. Nucleic acids of filaments of the pair (filaments 1 and 2) are the same. The different filaments of different pairs have different nucleic acids.

Organization of Specific Genes

Before I explain the new approach to protein and nucleic acid spirals biosynthesis in more detail, I will say that genes that produce polypeptides (1–8) of the protein spirals repeat along positive strands of ds DNA to produce groups of polypeptides that also repeat along polypeptide and protein spirals. Prokaryotic and eukaryotic cells also have different groups of genes that repeat along positive strands of ds

DNA; they are responsible for the phenotype variations of prokaryotic cells and differentiation of stem cells.

Prokaryotic cells comprise different filaments organized in different pairs that interchange parts of protein and nucleic acids spirals and have very similar ds DNA. Pairs of filaments have strings with different ds DNA. Each filament with its own ds DNA increases twice during the life of prokaryotic cells. The process of doubling of filaments begins immediately after binary fission and finishes with binary fission.

Groups of genes of positive strands of ds DNA of strings translate into new polypeptides and transform into tRNA. The new polypeptides synthesize genes that are similar to translated genes. Other groups of genes of positive strands of DNA of strings are only copied without translation into polypeptides. Negative strands are only copied.

String of filaments consist of synthetic and decomposition protein spirals with inside polypeptide spirals. Polypeptide spirals of synthetic or decomposition protein spirals consist of the eight polypeptides that make NAG and NAM according to the model. The eight genes must produce eight polypeptides of synthetic or decomposition protein spirals; the sixteen different genes of a group must exist along ds DNA to produce the sixteen polypeptides of synthetic and decomposition protein spirals.

1	1'	2	2'	3	3'	4	4'	5	5'	6	6'	7	7'	8	8'

The eight specific genes of decomposition protein spirals permeate the eight specific genes of synthetic protein spirals along positive strands of ds DNA.

The ds DNAs of strings of the pair of filaments are almost identical because strings interchange parts of protein spirals with nucleic acids according to the model. The ds DNA of strings produces separately decomposition and synthetic protein spirals in decomposition and synthetic strings. The sixteen genes must permeate each other in pairs of similar genes in any group of genes to produce polypeptides for synthetic and decomposition protein spirals. Any other combination cannot produce NAG and NAM and separate polypeptides for synthetic and decomposition protein spirals.

Some prokaryotic cells can produce proteins that help them adapt to environmental conditions; they have nucleic acids that produce pairs of the filaments. To transform into other pairs of the filaments, they must have other groups of genes that also repeat along the nucleic acids.

1	2	3	1	2	3	1	2	3

This hypothetical example shows three groups of pairs of similar genes that repeat along the positive strand of ds DNA. The genes can produce three groups of polypeptides. The three groups of the pairs of the similar genes permeate each other along the positive strands of ds DNA.

Different prokaryotic cells have up to several thousand different genes; the conclusion is that different pairs of different filaments with different nucleic acids with different groups of pairs of similar genes exist according to the new approach. Different nucleic acids of strings have different quantities of groups of pairs of similar genes that produce different lengths of strings and filaments also according to the new approach. This concept will be clear after explanation of biosynthesis of protein and the nucleic acid spirals.

New Approach to Biosynthesis of Protein and Nucleic Acid Spirals

Biosynthesis of protein and nucleic acid spirals is explained here differently from the way it is explained in the literature. The new approach to biosynthesis of protein and nucleic acids spirals cannot be explained without each other because they depend on each other. The most important components of structures of filaments will be repeated due to a better understanding of the new approach to protein and nucleic acid synthesis. The previous scheme of filament 1 is also used for better explanations of these phenomena.

Synthetic and decomposition strings of filaments 1 and 2 of a pair have very similar ds DNA regarding their quantity and quality. Interchange of parts of protein and nucleic acid spirals among synthetic and the decomposition strings of filaments 1 and 2 of the same pair causes the strings to have similar ds DNA.

The ds DNA of the strings' helicoid structure has complementary negative and positive strands. The helicoid structure of ds DNA is partially split at the coil parts due to partial spreading of protein spirals there and attraction of protein spirals. This partial splitting occurs during polarization and stops during depolarization.

Split positive and negative strands of ds DNA retain helicoid structures because the helicoid and spread structures of spirals attract them. They do not unwind according to the new approach because they can perform their functions without that. According to the literature, ds DNA unwinds and separates strands of ds DNA by different enzymes. The enzymes come and go without any logical explanations of how, where, and why. The new approach to the biosynthesis of the protein and the nucleic acid spiral is a logical explanation associated with different phenomena of cells and does not contradict them.

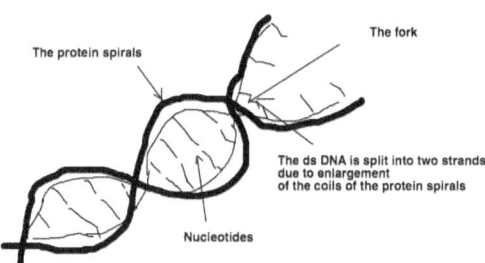

Ds DNA of the rope structure is partially split into negative and positive strands that retain helicoid structure.

Synthetic string of filament 1

>>> Synthetic protein spiral with ds DNA, tRNA, ss DNA of ds DNA >>>

<<< (New synthetic protein spiral with ds DNA and tRNA) + Decomposition protein spiral with ss RNA or ssDNA of dsDNA <<<

Coil part   Straight part

Decomposition string of filament 1

> <<< Decomposition protein spiral with ds DNA, tRNA, ss DNA of ds DNA <<<

> \>>> Synthetic protein spiral with ss RNA + (New decomposition protein spiral with ss RNA or ss DNA of ds DNA) \>>>

Coil part   Straight part

Synthetic protein spirals of synthetic strings and decomposition protein spirals of decomposition strings attract negative strands of ds DNA. According to the new approach, they function like DNA polymerase together with primers, positive strands of ds DNA broken during depolarization at the fork of the coil parts. A fork is where splitting ds DNA occurs. Negative DNA strands from straight to coil parts make positive DNA strands from their 5' to 3'. The synthesis of ds DNA on the positive strands is more complex because the positive strands are involved in biosynthesis of the protein spirals.

Decomposition protein spirals of synthetic strings and synthetic protein spirals of decomposition strings attract positive strands of ds DNA. Positive strands of ds DNA are broken at the fork of the coil parts and transformed into mRNA by the influence of the enzymes of protein spirals.

Protein spirals of strings function like RNA polymerase with primers that are positive strands of Okazaki fragments, which are synthesized in new synthesized specific polypeptides in a specific way according to the new approach. The mRNA attracts specific amino acids and triplets of tRNA and form new specific polypeptides and new ss RNA. The formation of polypeptides and new ss RNA from triplets of tRNA breaks attracted tRNA on the nucleotides. The ss RNA attracts complementary nucleotides and transforms into ds RNA pieces that are Okazaki fragments.

New synthesized specific polypeptides cut mRNA; those pieces as genes are used to make new specific polypeptides transform into tRNA, whereas other pieces of mRNA fuse with positive strands of ds

RNA pieces into incomplete ds RNA by help of the ligase enzymes that belong to decomposition protein spirals of synthetic strings or synthetic protein spirals of decomposition strings.

RNA polymerase synthesizes the gaps fragments among the Okazaki fragments to form complete ds RNA that will be transformed into ds DNA. The gaps fragments are negative pieces of ss RNA that are synthesized from their 5' to their 3'. This above hypothesis explains formations of the double amounts of the nucleic acids as ds DNA.

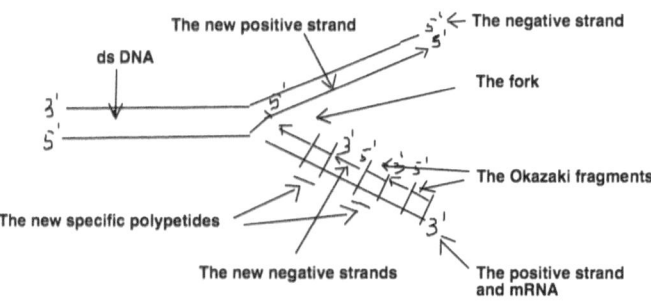

Biosynthesis of the double amounts of ds DNA

The new approach to biosynthesis of the protein spirals is different in many details from the central dogma of protein synthesis. Decomposition protein spirals of synthetic strings and synthetic protein spirals of decomposition strings attract positive strands of ds DNA. They break and transform positive strands of ds DNA into mRNA under influence of their enzymes. The difference between DNA and RNA is in their sugars (deoxyribose for DNA and ribose for RNA) and the nucleotide bases (adenine, thymine, guanine and cytosine for DNA and adenine, uracyl [instead of thymine], guanine, and cytosine for RNA).

After the splitting of ds DNA, its positive strands are exposed to NAD, ATP, and folic acid coenzymes of protein spirals that attract and transform them into RNA positive strands or mRNA. Deoxyribose of DNA is transformed into ribose of RNA by NAD coenzymes that take hydrogen atoms from their C2' and by ATP coenzymes that give hydroxyl groups to C2' during polarization.

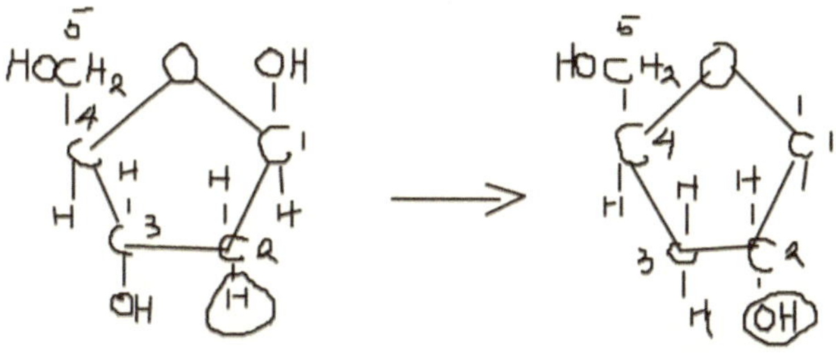

Deoxyribose   **Ribose**

Thymine molecules of DNA are transformed into uracil of RNA by folic coenzymes that take the methyl groups from thymine also during polarization.

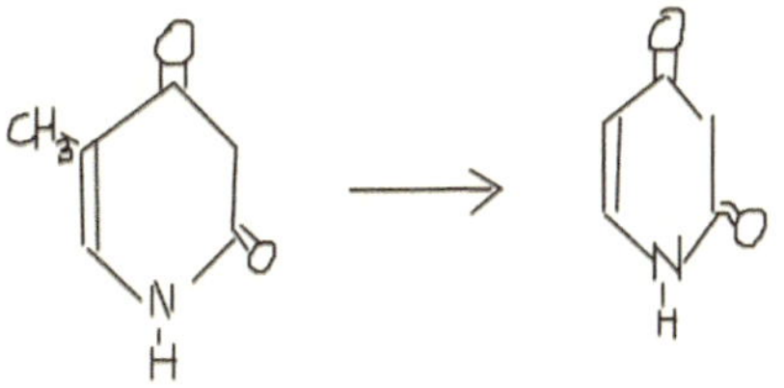

Thymine   Uracil

Enzymes change DNA into RNA and vice versa; this way is more natural than transcription of mRNA from unwound DNA according to the central dogma.

Genes of transformed DNA spirals in mRNA spirals are attracted by polypeptides of synthetic protein spirals of decomposition strings or by polypeptides of decomposition protein spirals of synthetic strings. Polypeptides attract genes that produced them; those genes free synthetic genes as mRNA pieces at synthetic strings or free decomposition genes as mRNA pieces at decomposition strings. Synthetic and decomposition genes permeate each other as pairs organized in groups that repeat along positive strands of ds DNA. This was already explained in the organization of the specific genes and their groups.

Synthetic protein spirals of synthetic strings or decomposition protein spirals of decomposition strings bring tRNA with specific amino acids. The tRNA has an amino acid attachment site and a

103

template recognition site according to the literature. The site is the anticodon of three bases that are complementary to the codon of mRNA also according to the literature. The ends of specific polypeptides of synthetic or decomposition protein spirals by their amino acids attract the 3' ends of tRNA or the amino acid attachment site also according to the literature. The function of the specific parts of tRNA is also different according to the new hypothesis that harmonizes the different phenomena.

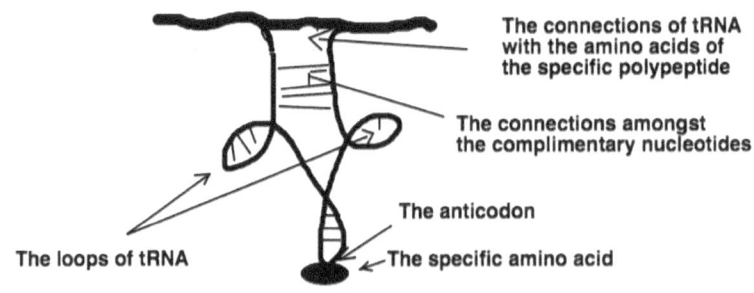

The connections of tRNA with the amino acids of the specific polypeptide

The connections amongst the complimentary nucleotides

The anticodon

The loops of tRNA

The specific amino acid

The structure of tRNA according to the new approach

According to the new approach, the new model of tRNA is similar to the model of tRNA according to the literature but with several changes. The anticodon of tRNA is a specific amino acid attachment site, a place of connection of tRNA with specific polypeptides according to the new approach. According to the literature, the specific amino acid attachment site of a tRNA does not attract specific amino acids but amino acids of specific polypeptides of protein spirals of synthetic or decomposition spirals.

The anticodon of tRNA (the triplet of the specific nucleotides) attracts a specific amino acid from B_6 coenzymes of close protein spirals. The codons of free genes of mRNA have specific triplets of nucleotides that also attract amino acids attracted to anticodons of tRNA according to the new approach. Codons of mRNA and anticodons of tRNA are complementary to attract each other as described in the literature.

The complimentary
nucleotides attract
each other

The reason for the new approach is that dogma of protein biosynthesis cannot logically explain polypeptide and nucleic acid biosynthesis connected with mRNA, tRNA, and other cell phenomena. Ribosomes are where new proteins are born according to the literature. Big and small units of ribosomes exist as groups of different proteins and nucleic acids in living cells according to the literature though many facts about them are explored and found in dead cells. The function of different proteins of ribosomes is not clear also according to the literature. Ribosomes as groups of different and coagulated proteins cannot exist in living cells because coagulated proteins cannot perform any function. The new structures of cells can explain the structure and the function of ribosomes in living cells because ribosomes are connected and explained with different phenomena of cells according to the new approach.

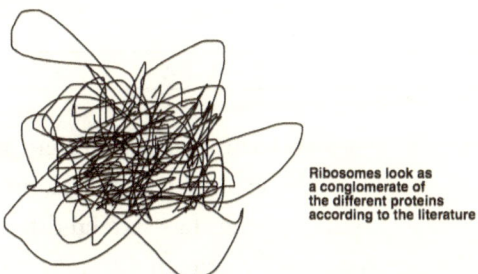

Ribosomes look as
a conglomerate of
the different proteins
according to the literature

Ribosomes are mRNA and amino acids attracted by tRNA during the making of polypeptides; mRNA and tRNA are attracted by opposite movement of protein spirals. The attraction is short lived according to the new approach.

The tRNAs attracted by synthetic protein spirals of synthetic strings move opposite to the movement of RNA (mRNA) attracted by decomposition protein spirals also of synthetic strings. Also, tRNA attracted by decomposition protein spirals of decomposition strings move opposite to the movement of RNA (mRNA) attracted by synthetic protein spirals of decomposition strings. The opposite movement of the tRNA with amino acids and mRNA with free genes (decomposition or the synthetic genes) enables polypeptide and nucleic acids biosynthesis.

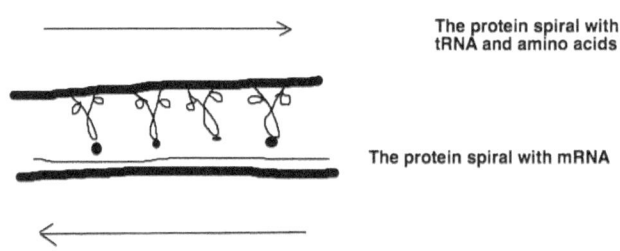

The protein spiral with tRNA and amino acids

The protein spiral with mRNA

The protein spirals with the specific nucleic acids move into the opposite directions.

Polypeptides of synthetic protein spirals attract synthetic genes of RNA and free decomposition genes as mRNA in decomposition strings. Also, polypeptides of decomposition protein spirals attract decomposition genes of RNA and free synthetic genes as mRNA in synthetic strings. Specific synthetic and decomposition genes are organized in pairs along positive strands of DNA or mRNA. There are eight pairs that make a group.

1	1'	2	2'	3	3'	4	4'	5	5'	6	6'	7	7'	8	8'

The eight specific genes of decomposition protein spirals permeate the eight specific genes of synthetic protein spirals along positive strands of ds DNA. A group of specific genes has sixteen different genes. The same and the different groups of the sixteen different genes repeat along the positive strand of ds DNA.

The free genes of mRNA attract specific amino acids attracted by tops of tRNA and synthesize new polypeptides during polarization according to the new approach. The new polypeptides attract coenzymes and transform into enzymes.

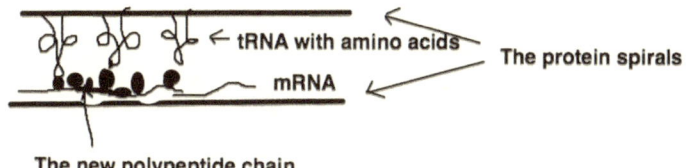

The diagram of the polypeptide synthesis

Newly synthesized specific polypeptides attract triplets of tRNA as ligase enzymes connect them in new ss RNA nucleic acid pieces. The rest of tRNA disassembles on nucleotides. The ss RNA attracts complementary nucleotides and makes ds RNA or Okazaki fragments. New specific polypeptides as restriction enzymes cut mRNA at the parts used for biosynthesis of polypeptides curl and transform into tRNA. The new specific polypeptides as ligase enzymes attract other parts of mRNA and connect them with positive strands of ds RNA in incomplete ds RNA.

New polypeptides as polymerase enzymes synthesize by the primers (the negative strands of the Okazaki fragments), the gap strands of incomplete ds RNA and transform it into a complete one. The new specific polypeptides attract coenzymes, mRNA, and new ss RNA or ds RNA simultaneously. This is explained by movements of the groups of amino acids around their axes.

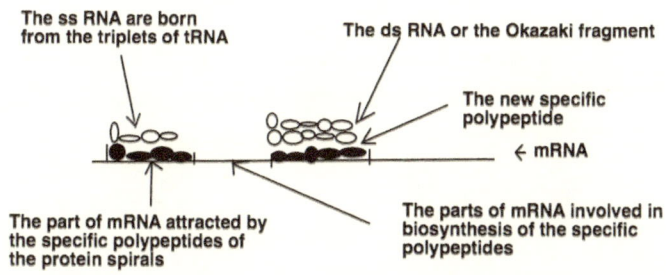

The atom groups (R1, R2, and others) of amino acids of polypeptides move fast around the axes of peptide bonds (CO, NH-) and attract nucleotides of the nucleic acids and coenzymes at almost the same time.

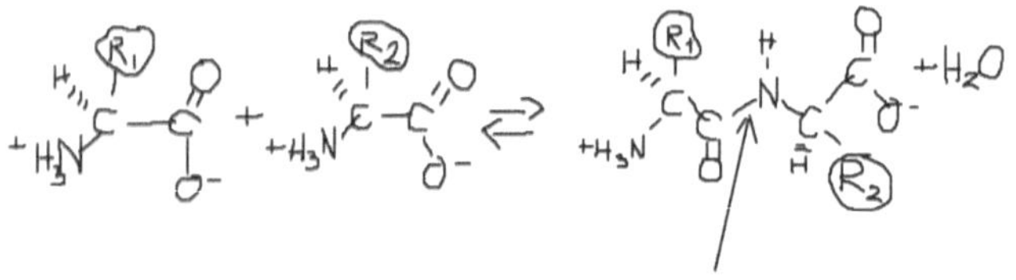

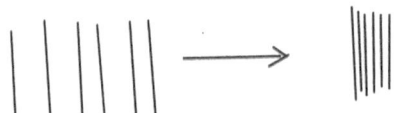

Peptide bond

The atom groups R1 and R2 of amino acids of polypeptides move very fast around peptide bonds.

The new polypeptides with tRNA, specific coenzymes, and specific ions attract each other forming new polypeptide spirals by their opposite electrical charges.

The new specific polypeptides attract each other by the opposite electric charges into the polypeptide spirals

The new polypeptide spirals attract new ds RNA with their ATP coenzymes and transform them into bigger dsDNA by influence of their enzymes during depolarization. ATP gets energy during depolarization and attracts hydroxyl groups from C2' of ribose. The free C2' attracts hydrogen ions released from NAD and FMN during depolarization and transform ribose molecules into deoxyribose ones.

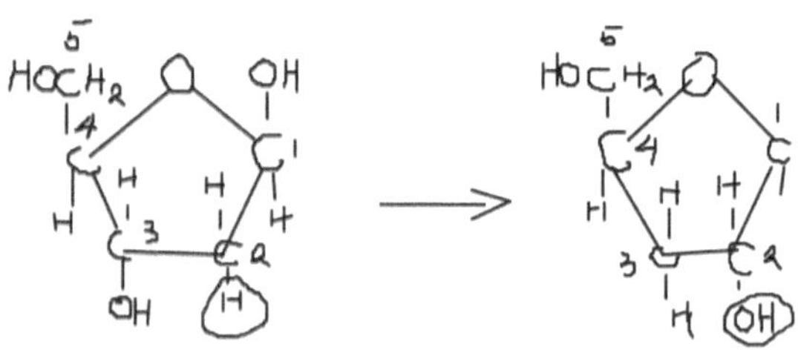

Deoxyribose

Ribose

Folic acid also releases methyl groups during depolarization. Uracil molecules in RNA attract methyl groups and transform them into thymine. The process of making DNA is the opposite of making RNA.

Thymine　　　　　　　　**Uracil**

The new polypeptide spirals curl and transform into new synthetic or decomposition protein spirals by attraction of opposite electrical charges of ions of polypeptides during depolarization. They also attract new ds DNA.

The polypeptide spiral transforms into the protein spiral during depolarization events

The new synthetic or decomposition protein spirals with new ds DNA break and spread at the overgrowth of the coil parts. Parts of new synthetic or decomposition protein spirals together with parts of new ds DNA and tRNA fuse with new synthetic or decomposition spirals and their ds DNA and tRNA by attraction among specific polypeptides during depolarization. Parts of new synthetic or decomposition protein spirals are broken at specific polypeptides at their ends. With specific polypeptides of parts of protein spirals, the opposite attracting ends enable fusion with protein spirals.

	Synthetic protein spiral	Overgrowth of synthetic protein spiral
Overgrowth of decomposition spiral with parts of new synthetic spirals protein and ds DNA spirals	Decomposition spiral	

Coil part   Straight part

	Decomposition protein spiral	Overgrowth of decomposition protein spiral
Overgrowth of synthetic spiral with parts of new decomposition and ds DNA spirals	Synthetic protein spiral	

Coil part   Straight part

The parts of new synthetic or decomposition protein spirals with parts of ds DNA fused with new synthetic or decomposition protein spirals and their nucleic acids move from the coil to the straight parts. They break into parts of synthetic or decomposition protein spirals with parts of ds DNA spirals at overgrowths of their straight parts also at their polypeptides. The parts of synthetic or decomposition protein spirals with parts of ds DNA spirals interchange at straight parts. Now, parts of synthetic protein spirals in decomposition strings move from straight to coil parts; parts of decomposition strings in synthetic strings move from straight to coil parts.

The strings have two ds DNA spirals that move in opposite directions; one is old while the other is new. The opposite moving protein spirals of strings attract old ds DNA spirals. The old ds DNA spirals move from straight to coil parts and enable biosynthesis of polypeptides, nucleic acid pieces, and new ds DNA from their negative strands.

The new ds DNA synthesized in the negative strands of the old ds DNA is attracted only by one protein spiral. Positive strands of old ds DNA are spent on biosynthesis of polypeptides and nucleic acids, which transform into parts of protein spirals and new ds DNA spirals during depolarization.

The parts of new ds DNA spirals fuse with new ds DNA spirals and move from coil to straight parts with them. Protein spirals that move from coil to straight parts attract new ds DNA spirals only for their positive strands. The new ds DNA spirals break into parts together with parts of specific protein spirals. Parts of new ds DNA spirals fuse with new ds DNA spirals. When old ds DNA spirals are spent, they are replaced with new ds DNA spirals that transform into old ds DNA. This occurs after divisions of filaments and binary fissions of prokaryotic cells.

The parts of protein spirals and ds DNA that are interchanged and fused among synthetic and decomposition strings of filaments enable biosynthesis of polypeptides and nucleic acids simultaneously and growth of protein spirals and strings.

Strings of filaments accumulate a double amount of protein and nucleic acids in filaments before their divisions and binary fission of a prokaryotic cell. Filaments do not change lengths though they have double amounts of proteins and nucleic acid spirals because they have organized strings that enable their growth and division into filaments.

Organization of filaments and the other structures enables growths of strings and divisions of filaments.

CHAPTER 12
BINARY FISSION AND DIFFERENT STRUCTURES OF PROKARYOTIC CELLS

Prokaryotic cells divide into two daughter cells almost identical to the mother cell. This division is known in the literature as binary fission, which is not explained at all in the literature due to poor understanding of the basic structures of prokaryotic cells and their connections with universal metabolism.

The new view of binary fission is different in comparison with confusing approaches of contemporary science as analyzed in the appendices. The new view must be also explored and proven despite its logic. The hypothetical structures and phenomena of prokaryotic cells explain binary fission and are in a causal relationship with it. Many hypothetical phenomena of prokaryotic cells are shortly repeated due to better understanding of binary fission.

Prokaryotic cells consist of filaments organized in pairs that interchange parts of protein and nucleic acid spirals. Filaments consist of two strings with specific organization. Interchange of protein spirals enables growth of filament strings in two opposite directions. A string has a straight part at one side and a coil part at the other. Another string of a filament has a straight part at one side that is with the coil part of the first string. The coil part of another string is also together with the straight part of the first string. This organization of filaments enables their growth and division by the phenomenon of chewing gum.

The model of a filament consists of the two opposite growing and resting strings. The string of filament has the coil part that grows and the straight part that does not grow. This structure of filaments enables a division of filament by the phenomenon of chewing gum.

Before I explain division of filament and binary fission in a cell, I will repeat some previous information to offer a better understanding of this phenomenon. The logic of the division of filaments created necessary structures of filaments for its division. The strings consist of synthetic and decomposition spirals that move in opposite directions by their polypeptide spirals, which attract each other by metabolites in filaments and polypeptide spirals of close strings of filaments. The attraction provides for all filaments of cells to merge in a system. The cytoplasm of prokaryotic cells looks like a compact mass in electronic images due to attraction among filaments but also due to their seeming coagulation.

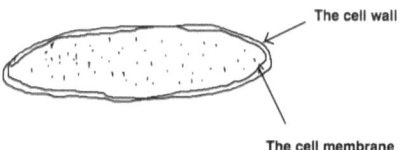

The cell wall

The cell membrane

All filaments merge into a compact structure by attractions of their polypeptides among themselves by their metabolites that act as inner glue. Electron images of bacteria do not show and explain the structures of the cell membrane, the cell wall, and other structures.

The two strings wind around each other by their opposite growth and movement; coil parts grow around straight parts that attract coil parts and do not allow coil parts to separate and overgrow them even during depolarization. The force of winding strings occurs in hungry coenzymes of opposite moving protein spirals due to inherent energy that enables them to attract ions from metabolites.

This winding and accumulation of potential energy and torsion causes a division of filaments. Filaments have synchronized growth that enables them to undergo torsion and cutting at the similar time that causes binary fission of a prokaryotic cell.

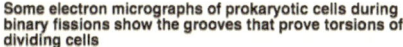

Some electron micrographs of prokaryotic cells during binary fissions show the grooves that prove torsions of dividing cells

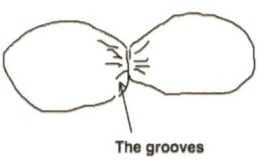

The grooves

Binary fission of a prokaryotic cell is due to the synchronized cutting torsions of its filaments. Many prokaryotic cells have the division septum during binary fission due to the pressure of uninterrupted peptidoglycan layers of the cell wall.

Immediately after division, cut filaments of daughter cells have highly twisted coil parts around the straight parts; this causes unwinding during depolarization due to energy in the highly twisted coils. This unwinding occurs in the direction of the missing straight parts of strings. They make new straight parts of renewed filaments during depolarization.

All these phenomena happen quickly. Renewed filaments have single strings on one side and two strings on the other. Pairs of renewed filaments of daughter cells interchange parts of protein spirals the same as the predecessor filaments of mother cells. Straight strings of renewed filaments grow around single strings. The straight parts grow because they have highly coiled nucleic acid spirals that quickly produce protein spirals.

The life span of a prokaryotic cell occurs in three phases: unwinding, immature, and winding (mature) one. The unwinding phase occurs immediately after binary fissions of prokaryotic cells and lasts a very short time. Coils twisted around straight ones of cut filaments transform into single strings. Pairs of renewed filaments with single strings regenerate into mature filaments with two strings.

The immature phase comes immediately after the unwinding phase; it lasts longer than the unwinding phase and shorter than the winding phase. Straight parts of strings grow around single strings during this phase. The immature phase finishes when strings growing around straight strings reach their ends.

The winding phase starts when growing parts of strings around straight ones reach their ends. Straight parts of the strings grow also around the single strings and produce new coil parts of strings during this phase. The coils become very twisted during this phase, which causes thickening of filaments, torsion, and cutting. The winding phase finishes when filaments are cut.

The torsion of thick filaments at their middles resembles the torsion of chewing gum being pulled apart. Therefore, this phenomenon is called the phenomenon of chewing gum.

Different Structures of Prokaryotic Cells

Different filaments continuously attract each other during the life span of prokaryotic cells with different levels of attractions during polarization and depolarization. The attracted filaments build different structures of prokaryotic cells, bodies of different shapes (cocci, bacilli, spirals, and others), cell membranes, and the walls of gram-negative and gram-positive prokaryotic cells (peptidoglycan, polysaccharides, and proteins), flagella, pili, cytoplasm, ribosomes, plasmids, and division septum.

Contemporary microbiology explains the structure of prokaryotic cells as separate entities without deeper connections among them. According to the new approach, filaments can explain the different structures of prokaryotic cells. Filaments are made of strings with nucleic acid and protein spirals that move in opposite directions during metabolism. This approach negates circular chromosomes. The model of a filament has two strings with two protein spirals with polypeptide spirals inside.

Each string of filaments has nucleic acids that exist between protein spirals. Chromosomes do not exist as circular structures due to reasons explained (see appendix III). The ds DNA of strings of filaments can have one or more groups of pairs of similar genes that repeat on them.

1	2	3	1	2	3	1	2	3

This hypothetical example shows three groups of pairs of similar genes that repeat along the positive strand of ds DNA.

Any group of the pairs of similar genes has sixteen genes grouped in eight pairs.

1	1'	2	2'	3	3'	4	4'	5	5'	6	6'	7	7'	8	8'

The specific group of the pairs of the similar genes repeats along ds DNA.

Protein spirals that move from straight to coil parts of filaments attract and move positive strands of old ds DNA and old ds DNA though the negative strands of old ds DNA are attracted by the protein spirals that move in opposite directions.

The ds DNA that moves from straight to coil parts is marked as old ds DNA whereas ds DNA that moves from coil to straight parts is marked as new ds DNA, which is attracted and moved by positive strands with protein spirals that move from coil to straight parts.

Parts of protein and new ds DNA are broken and expelled at the straight part as overgrowth and are attracted by straight parts of other strings of paired filaments. Parts of new ds DNA are attracted and fused by old ds DNA that moves to coil parts.

Old ds DNA is split into negative and positive strands at coil parts. Parts of positive strands of ds DNA are transformed into positive strands of RNA that contain mRNA.

Polypeptide biosynthesis occurs in mRNA, which transforms into tRNA attracted by the new polypeptides of the new polypeptide spirals. The new polypeptides of new protein spirals synthesize new ss RNA that transform into new ds RNA. New ds RNA fuses with ss RNA of attracted positive parts of ss RNA by specific polypeptides and transforms into incomplete ds RNA, which transforms into complete ds RNA, which transforms into new ds DNA. The new ds DNA fuses with the rest of the new ds DNA and moves from coil parts to straight ones with new parts of protein spirals. The new ds DNA is expelled from filaments as the part of the protein and nucleic acid spirals.

The ds DNA produced by the negative strands of ds DNA activates after the division of filaments and transforms into old DNA. Negative strands of old DNA produce half the necessary nucleic acids before division of strings, whereas positive strands of old ds DNA produce the other half. The strings produce double the amount of protein and nucleic acid spirals before their divisions in filaments.

Existing phenomena of transformation, conjugation, transduction, and jumping genes prove the new approach. These phenomena cannot exist if the chromosomes of prokaryotic cells are circular because circular chromosomes have closed ds DNA. The strings of filaments with their own nucleic acids contradict the contemporary model of circular chromosome of prokaryotic cells.

Protein spirals with tRNA and their amino acids move opposite from the protein spirals that bring mRNA as explained. Protein spirals with mRNA attract amino acids attracted by tRNA and form a ribosome. Protein spirals with mRNA and amino acids make small units of ribosomes, whereas protein spirals with tRNA and the same amino acids make big units of ribosomes.

A ribosome or a group of ribosomes (polyribosomes) lasts for a very short time during attraction of specific amino acids for specific triplets of mRNA and disappear immediately after attracting amino acids.

The ribosomes described here are not the same as those described in the literature. Proteins of the new ribosomes are not connected only with ribosomes as is explained in the literature. The two opposite protein spirals build many ribosomes that do not clump, but ribosomes look like balls in the literature because they were seen under electron microscopes and explored by different methods as coagulated structures in dead cells.

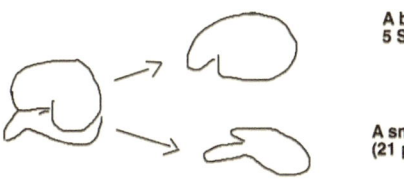

A big unite has 23 S and
5 S r RNA (34 proteins)

A small unite has 16 S rRNA
(21 proteins)

A ribosome of prokaryotic
cells

A ribosome consists of a big and a small unite according to the literature. The units appear as coagulated proteins.

Big concentrations of tRNA appear at the coil parts, and their concentration gradually falls to the straight parts of the strings because tRNA spends and disassembles along the strings during biosynthesis of polypeptides. The new view of ribosomes and protein biosynthesis is different from the structure of ribosomes and the central dogma of polypeptide and nucleic acid biosynthesis.

Septum appears during binary fission of prokaryotic cells without any explanation of its origin and connections with other cell structures in contemporary literature. Septum appears during division of some prokaryotic cells due to pressure of unbroken peptidoglycan and polysaccharides molecules on a dividing prokaryotic cell during binary fissions according to the new approach. The septum of prokaryotic cells consists of broken and unbroken filaments.

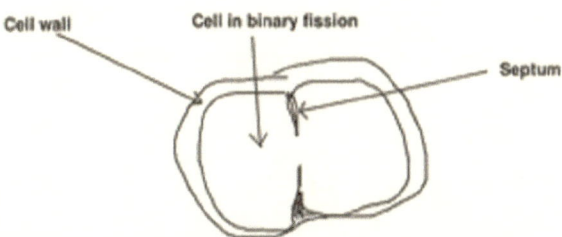

Cell wall   Cell in binary fission   Septum

PRESSURE OF CELL WALL AROUND DIVIDING A PROKARYOTIC CELL CAUSES
APPEARANCE OF SEPTUM WHICH CONSISTS FROM CELL MEMBRANE

Filaments attracted between the environment and the cell cytoplasm make plasma membranes or cell membranes. Plasma membranes are not compact sandwich membranes with embedded proteins that act as doors for transport of different metabolites as described in the literature. Plasma membranes are made up of attracted filaments with continuous movement. Strings of filaments also contain fat and fatty acids. Filaments of plasma membrane are also attracted with filaments of cytoplasm with which they share metabolites for synthesis and decomposition.

Filaments of cytoplasm and plasma membrane produce polysaccharides, peptidoglycan, and many other molecules as a unified system. These molecules are exported outside the cells as peptidoglycan, polysaccharides, and others. The cell wall makes a rigid wall due to polysaccharides and peptidoglycan molecules that do not move out of the cell membrane though they are continuously decomposed and synthesized.

Cell walls protect cells from the environment and feed cells by decomposition. Cell walls act as elastic socks that shape cells including the division septum during their binary fission. Strings of filaments and filaments have the greatest concentrations of nucleic acids at the ends or on straight parts. Each string of filaments has nucleic acids, without circular chromosomes that do not exist in prokaryotic cells.

Some prokaryotic cells including mycoplasma spp. have different shapes and do not have cell walls though they have plasma membranes because they do not produce complex polysaccharides and peptidoglycan that press their cytoplasm and make stable shapes. They are known as prokaryotic cells without cell walls.

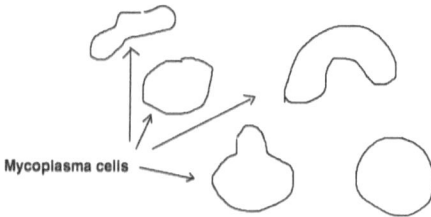

Mycoplasma spp. has different shapes of bodies.

Different filaments get out of prokaryotic cells and make flagella and fimbriae or pili. They have the same structure as filaments do.

Fimbriae and flagella differ in diameter, size, and quality of proteins. Flagella are longer and thicker than fimbriae are, and they both have the same functions as other filaments—they release, attract, and interchange protein spirals and new ds DNA by straight parts of strings.

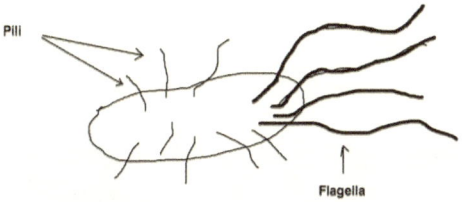

Flagella and fimbriae are different filaments that get out from bodies of prokaryotic cells according to the new approach.

Prokaryotic cells of the same species can interchange protein spirals, new ds DNA, and other molecules among themselves by flagella and fimbriae whereas prokaryotic cells of different species cannot due to differing protein spirals. Fimbriae of the same species enable the interchange of resistance genes to different antimicrobial agents. The interchange of parts of protein spirals with parts of new ds DNA and other molecules among flagella and fimbriae outside prokaryotic cells makes accumulations known in the literature as biofilm. That helps them survive in differing environments during their short lives. Biofilm provides them constant moisture, optimal temperature, food, and protection against the immune system and other organisms.

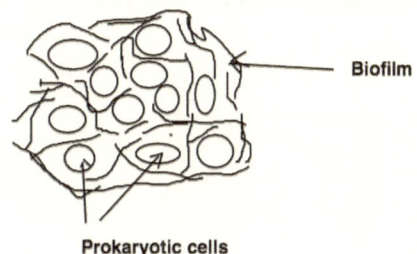

Many prokaryotic cells produce biofilm that consists of protein spirals, polysaccharides, minerals, and others. Biofilm is a product of the interchange of the parts of protein spirals and nucleic acids among prokaryotic cells of the same species. Biofilm provides them with water, food, and protection.

Cell membranes consist of attracted filaments that give the impression of flat or sandwich structures with embedded proteins in electron microscope pictures.

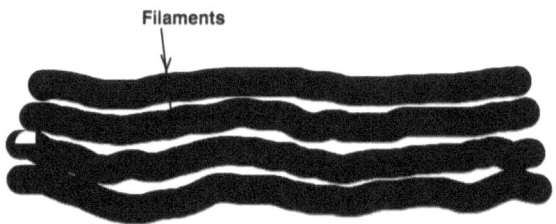

Filaments on the periphery of prokaryotic cells make the cell membrane

The plasma membrane of prokaryotic cells attracts parts of protein spirals and new ds DNA released by flagella and fimbriae with different intensities. If plasma membranes strongly attract parts of protein spirals, they produce cell walls of gram-negative prokaryotic cells. The parts of the protein spirals strongly attracted by plasma membranes are not easily drawn into the cytoplasm and utilized among filaments.

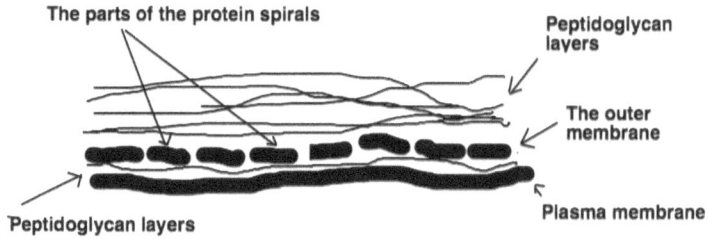

The cell wall of Gram-negative prokaryotic cells

If the plasma membrane does not attract parts of protein spirals, it produces cell walls of gram-positive prokaryotic cells.

Peptidoglycan layers

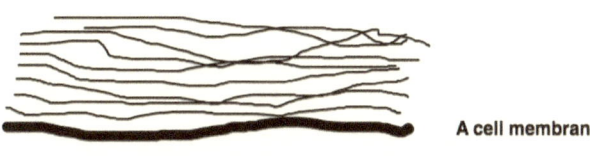

A cell membrane

Gram-positive prokaryotic cells

Many filaments of prokaryotic cells do not get out of the cells as fimbriae and flagella. They bend at their ends and stay in the cytoplasm; they release parts of the protein spirals and nucleic acids in the cytoplasm and interchange them among themselves. The accumulation among filaments in prokaryotic cells appears as light areas seen in the electronic images. The light areas are known as nucleoids according to the literature.

Nucleoid   A plasma membrane

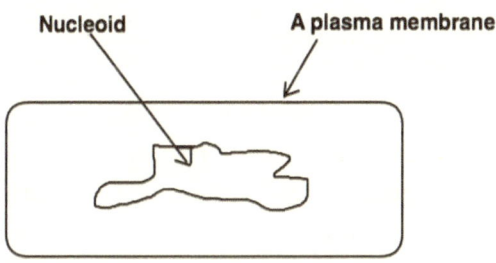

Pairs of filaments release parts of protein spirals and nucleic in the cytoplasm. They interchange them among themselves. Accumulation of parts of protein spirals and nucleic acid at prokaryotic cells are seen as the light areas or the nucleoids at the electronic images.

Some pairs of filaments do not get the necessary parts of protein and the nucleic acid spirals for their growth due to weak attraction of parts of protein spirals and nucleic acids at the straight parts of the strings and for other reasons. They cannot grow though they synthesize protein and nucleic acid spirals. They become smaller by releasing parts of protein and nucleic acid spirals due to their poor interchange among the straight parts of the strings. The results are small pairs of filaments called plasmids in the literature.

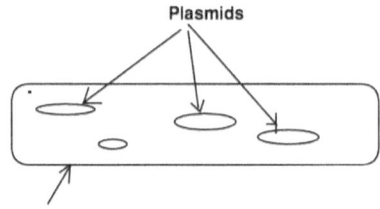

A plasma membrane

Plasmids are not small and circular DNA molecules in living prokaryotic cells as they appear in electronic images. Plasmids are short pairs of filaments that interchange parts of protein and nucleic acids spirals in living cells according to the new approach. The ends of ds DNA of the short filaments attract each other in the circular ds DNA in the dead prokaryotic cells. If the plasmids are circular ds DNA, they cannot interchange parts of protein and nucleic acid spirals with themselves and other cells.

Different bodies and sizes of prokaryotic cells contain different numbers of pairs of filaments with different lengths and qualities, which contain different lengths of polypeptides in polypeptide and protein spirals. Different lengths contain different concentrations of coenzymes and metabolites. All these characteristics order the different shapes of prokaryotic species cells.

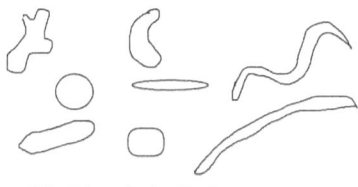

Different shapes of prokaryotic cells

Different bodies and sizes of different prokaryotic cells contain a different number of pairs of filaments with different lengths and qualities.

CHAPTER 13
DIFFERENT PROKARYOTIC CELLS AND THE MAJOR GROUPS

Prokaryotic cells have different numbers of filaments with different lengths and qualities. Filaments of prokaryotic cells are organized in pairs that interchange parts of protein and nucleic acid spirals. These filaments have different sizes of polypeptides with different concentrations of coenzymes. The polypeptides attract different concentrations of metabolites in decomposition and synthetic protein spirals. Different concentrations of coenzymes and metabolites enable different decomposition and synthesis of molecules that produce characteristics of prokaryotic cells.

Three major groups of prokaryotic cells live together in different and blended ecosystems—the decomposition group, the synthetic group, and one between the extremes. Prokaryotic cells of the decomposition group have greater decomposition of complex molecules into simple ones than their synthesis. Prokaryotic cells of the synthetic group have greater synthesis of complex molecules from simple ones than do decomposition groups. Prokaryotic cells of the middle group are between the two and can be slightly on the decomposition or the synthetic side. They produce more or less the same amounts of the simple and complex metabolites.

Prokaryotic cells of the decomposition group provide energy mainly by decomposition of complex molecules, which release energy in their molecular bonds. Complex molecules are the product of synthetic groups of prokaryotic cells and other synthetic organisms. The prokaryotic cells of the decomposition group produce porphyrin-Fe++ complexes that attract oxygen or metabolites from the Krebs cycle; these complexes neutralize hydrogen atoms with oxygen atoms or metabolites from the Krebs cycle during depolarization.

Prokaryotic cells of the synthetic group provide energy mainly from the Sun, thermal water, and other sources; they receive some energy by decomposition of complex compounds. The majority are photosynthesizers that capture the Sun's energy by porphyrin-Mg++ complexes that attract simple molecules (CO_2, N_2, SO_2 and others). They split them into atoms ($CO_2 \rightarrow C + 2O$, $N_2 \rightarrow 2N$, $SO_2 \rightarrow S + 2O$). The specific atoms neutralize hydrogen ions ($C + 4H \rightarrow CH_4$, $O + 2H \rightarrow H_2O$, $2N + 6H \rightarrow 2NH_3$, $S + 2H \rightarrow H_2S$ and others). The specific porphyrin-Mg++ complexes transfer energy to AMP that transforms into ATP (appendix XVII).

The polypeptide spirals of synthetic prokaryotic cells produce porphyrin molecules similar to chlorophyll from porphobilinogens on their biotin coenzymes by Mg++ during depolarization. The unknown coenzymes attract Mg++ ions and release them during depolarization. Porphyrin-Mg++ complexes similar to chlorophyll attract the Sun's energy and transfer it to AMP also during depolarization. They disassemble after transferring energy to AMP that transforms into ATP. New cycles of polarization start after their dissembling and by the energy of ATP.

Chlorophyll a

Prokaryotic cells have porphyrins similar to chlorophyll a.

The specific porphyrin Mg++ complexes of synthetic groups can split oxygen and other molecules to perform oxidation of hydrogen atoms when synthetic prokaryotic cells do not have the Sun's energy at night; they decompose complex molecules and use their energy.

Prokaryotic cells of decomposition groups have porphyrin-Fe^{++} that cannot attract energy from the Sun or other sources, so they derive energy by decomposition of complex molecules. Decomposition prokaryotic cells live with synthetic ones in the same ecosystems to get the complex molecules.

The prokaryotic cells of the middle group have most likely the different porphyrin Fe^{++} and Mg^{++} complexes. They can attract the Sun's energy and molecules (CO_2, N_2, SO_2, and others). They also attract and split complex molecules.

The three major groups of prokaryotic cells synthesize and decompose the same basic metabolites with the Krebs cycle.

The Phenotype Variation

Many prokaryotic cells can change phenotypes based on environmental conditions; this is called phenotype variation, and it is due to different pairs of similar genes along the strings' nucleic acids. Different pairs of similar genes enable this phenomenon in harmony with other phenomena (appendix XXXI).

1	2	3	1	2	3	1	2	3

This hypothetical example shows the three different groups of the pairs of the similar genes that repeat along the positive strands of the ds DNA. Each group has eight pairs of the specific genes or the sixteen genes. The specific groups of the pairs of the similar genes are important because they enable prokaryotic cells the phenotype variations under the different environmental conditions and better surviving.

Prokaryotic cells live in groups to better survive under different environmental conditions as will be explained. Prokaryotic cells of a species must have the same major characteristics according to the old and new definition of prokaryotic species. Strains are similar subgroups in the same species; they have the same major characteristics of their species but differ in minor characteristics also according to the old and new definitions. The new definition of prokaryotic species and their strains will be explained in chapter 15.

Prokaryotic cells of a strain change very slowly after binary fissions in different environment conditions over a long time. Accumulation of minor changes in a strain changes its major characteristics and produces new major characteristics of a new species. This transformation takes a long time during their evolution.

The transformation of one species into another is recorded in their nucleic acids as different groups of gene pairs. The hypothetical origin of these groups that repeat along the ds DNA of the strings at different prokaryotic species is explained in the next chapter.

Archaebacteria are prokaryotic cells that did not change drastically due to stable environmental conditions. They have few groups of gene pairs that exist at the strings of the pairs of filaments and a small number of different genes. Other prokaryotic cells changed a lot due to environmental conditions over a long time. They have many groups of gene pairs along nucleic acids at the strings.

The following scheme is an example of ds DNA of the string that has only two groups of gene pairs that repeat in ds DNA of strings of the pair of filaments. Each group of pairs of similar genes has synthetic and decomposition genes as two subgroups that permeate each other. They work as separate subgroups during biosynthesis of synthetic and decomposition protein spirals at synthetic and decomposition strings of the pair of filaments.

1	1'	2	2'	3	3'	4	4'	5	5'	6	6'	7	7'	8	8'

A group of the pairs of the similar genes of filament has eight pairs of specific genes that permeate each other and are responsible for production of synthetic and decomposition protein spirals.

Gene pairs at different filaments of a prokaryotic cell are active only under specific environmental conditions. They produce the phenotype that has the best chance of survival under specific environmental conditions. The active specific group changes to another active specific group of genes under specific environmental conditions, which cause this change or jump from the active group of genes to another.

The other active group of genes produces the other phenotype that has the best chance of surviving under specific environmental conditions. The environment can produce different phenotypes or variation in some prokaryotic species only if prokaryotic cells have different groups for varying environmental conditions.

The phenomenon of the phenotype variation can be explained this way. Synthetic and decomposition genes of the same group of gene pairs are not attracted by polypeptides of synthetic and decomposition spirals under other environmental conditions because their coenzymes do not attract them due to other environmental conditions.

Polypeptides of synthetic and decomposition spirals attract synthetic and decomposition genes from the other group of gene pairs according to specific environmental conditions. The attraction of the specific polypeptides of the specific protein spirals for other groups of genes produces the phenomenon of phenotype variation. The phenomenon of phenotype variation enables different species of prokaryotic cells to survive in different environmental conditions. Some groups of prokaryotic cells are not used at all because certain environmental conditions no longer exist. These genes are documents of their evolutionary history.

CHAPTER 14
HYPOTHETICAL EVOLUTION OF PROKARYOTIC CELLS AND UNIVERSAL LAWS OF NATURE

The first forms of life appeared most likely in an optimal environment a long time ago (appendix XL). The optimum conditions for life had been these.

- water with optimum temperature
- optimum salinity and optimum pH for the best work of enzymes
- dissolved oxygen and different molecules with carbon, nitrogen, sulfur, and other atoms
- stable source of energy—the Sun

Evolution of simple molecules such as amino acids, simple sugars, and others that appear during metabolism of fructose molecules necessary for life was very slow under the influence of unknown natural forces. The first forms of life started with the appearance of these molecules, which attracted each other by unknown natural forces. They transformed into archaic short polypeptides with nucleotides and coenzymes that attracted simple molecules.

Simple molecules appeared later as metabolites during metabolism. Nucleotides transformed into nucleic acid pieces in archaic short polypeptides, which with coenzymes and simple molecules transformed into archaic associations of archaic polypeptides. Those associations transformed later into protein spirals and archaic strings, filaments, and prokaryotic cells.

Archaic associations of polypeptides reproduced each other due to energy and environmental conditions. The interaction of coenzymes with simple metabolites had been present likely in archaic strings due to the complex organization that enabled their interaction. The interaction of coenzymes with simple metabolites synthesized polysaccharides, fatty acids, fats, and other complex molecules.

Archaic associations with polypeptides were most likely the first form of life and the predecessor of contemporary prokaryotic and eukaryotic cells. The hypothetical archaic forms of predecessors of prokaryotic cells are these:

1. Archaic polypeptides
2. Archaic protein spirals
3. Archaic strings
4. Archaic filaments
5. Archaic prokaryotic cells
6. Contemporary prokaryotic cells

Archaic Polypeptides

Amino acids joined most likely into archaic polypeptides by unknown forces of nature. Archaic polypeptides attracted predecessors of unknown coenzymes that attracted molecules necessary for transformation in coenzymes. Unknown coenzymes attracted simple molecules such as N_2, CH_4, CO_2, H_2S and later more complex molecules as dihydroxyacetone P, glyceraldehyde-3P, fructose, ribose, pyruvate, and other metabolites as ions. The complex molecules had been products of their interactions.

Amino acids of archaic polypeptides attracted predecessors of purine and pyrimidine molecules that transformed with other molecules into triplets of nucleotides. Triplets of purine and pyrimidine molecules attracted ribose and phosphate molecules forming short pieces of nucleic acids that as positive strands of nucleic acids attracted amino acids and synthesized new polypeptides. The new polypeptides attracted predecessors of unknown coenzymes that had attracted other molecules necessary for transformation into coenzymes.

The mother archaic polypeptides and the new archaic polypeptides repelled each other due to having the same electrical charges. Amino acids of the new archaic polypeptides attracted predecessors of purine and pyrimidine molecules that had transformed with other necessary molecules into triplets of nucleotides, which attracted ribose and phosphate molecules forming short pieces of new archaic nucleic acids. The cycle of synthesis of archaic polypeptides repeated.

The attraction of archaic polypeptides for coenzymes and triplets of nucleotides was possible due to movement of atom groups of amino acids with high speed around the axes of their peptide bonds. Below are pictured two amino acids with peptide bonds and the atom groups of amino acids among them.

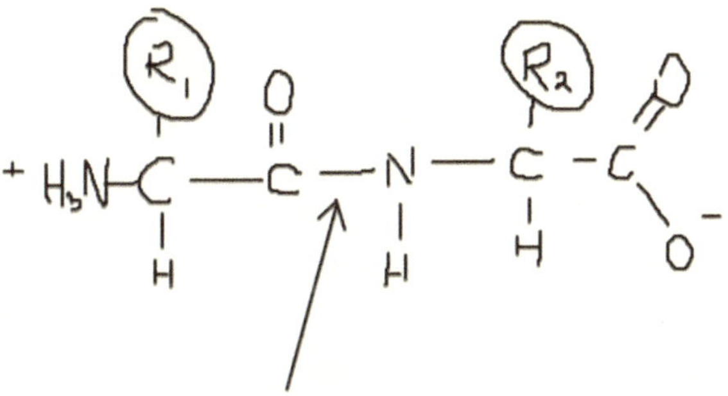

Peptide bond

Specific atom groups (R1, R2) of amino acids are between the peptide bond and move quickly around them.

Triplets of nucleotides of archaic positive strands of nucleic acids attracted amino acids of archaic polypeptides and free specific amino acids simultaneously. This duality of attraction was possible due to movement of their nucleotides with pentose sugars at high speed around the axes of their phosphate molecules. The picture below shows the strand of nucleic acids with nucleotides and sugars.

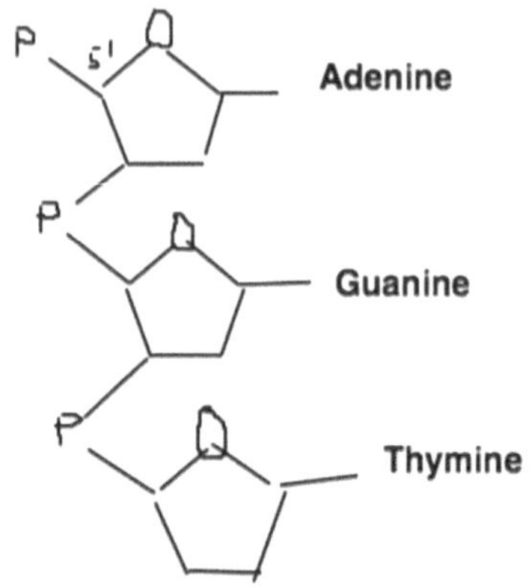

Adenine

Guanine

Thymine

Nucleotides and sugars move around phosphate axes of nucleic acid at high speed.

Archaic positive strands of nucleic acids did not have enough nucleotides to make complementary and archaic negative strands due to poor biosynthesis of nucleotides in archaic polypeptides. Attraction of archaic polypeptides for nucleotides of short nucleic acids and for coenzymes enabled primitive metabolism and reproduction. Attraction of short nucleic acids for amino acids of the same archaic polypeptides and for free amino acids took place almost simultaneously. It enabled reproduction of new archaic polypeptides on short nucleic acids also under the unknown environmental conditions. The new archaic polypeptides with attracted coenzymes and metabolites had repelled the mother archaic polypeptides due to the same electrical charges of their metabolites and had caused separation from their mothers.

Archaic polypeptides with metabolites lived in archaic associations in water under unknown environmental conditions. Archaic association means different forms of life with optimal environmental conditions that enable them to survive. They had attracted each other in archaic polypeptide spirals according to their attraction affinities. Archaic polypeptide spirals negated archaic polypeptides and made an evolutionary jump to further development of archaic protein spirals.

Archaic Protein Spirals

The archaic long polypeptide spirals had transformed into the archaic protein spirals

Short archaic polypeptide spirals attracted each other into archaic protein spirals according to the order of specific polypeptides attractions.

Archaic protein spirals with short pieces of nucleic acids in polypeptides lived in association and produced the same molecules as archaic polypeptides did. They did not synthesize polysaccharides, fatty acids, fats, and complex molecules because they did not have an organization of strings necessary for making complex molecules.

The short pieces of the archaic nucleic acids attracted complementary amino acids, triplets of nucleotides attracted amino acids and vice versa. Amino acids attracted each other by the influence of polypeptide spirals producing newly synthesized archaic polypeptides, which had attracted unknown coenzymes and their metabolites with electrical charges.

Due to their attraction, these polypeptides transformed into protein spirals that had repelled the mother protein spirals due to having the same electrical charges. Replication of archaic polypeptide spirals took place in archaic association.

Researchers (undergraduate Jason Lang, graduate students Yugang Bai and Hua Lu, and materials science and engineering professor

Jianjun Cheng) found that elongating side chains with charged ends enabled short proteins to coil into a stable helix.[126]

Close relationships between protein spirals and their opposite electrical charges caused attraction of protein spirals in archaic strings under unknown environmental conditions. Archaic strings had the same organization of protein spirals. Archaic protein spirals in archaic strings had moved each other in opposite directions performing metabolism of metabolites.

Archaic Strings

Polypeptides of two protein spirals had opposite concentrations of unknown coenzymes and ions that had attracted each other into unstable forms of archaic short strings. They made the unity and the conflict of opposites. Archaic strings negated archaic protein spirals and had made the evolutionary jump to further development of archaic filaments and complex cylinders. The archaic polypeptides, protein spirals, and strings were predecessors of archaic prokaryotic and eukaryotic cells.

The protein spirals consist from the polypeptide spirals

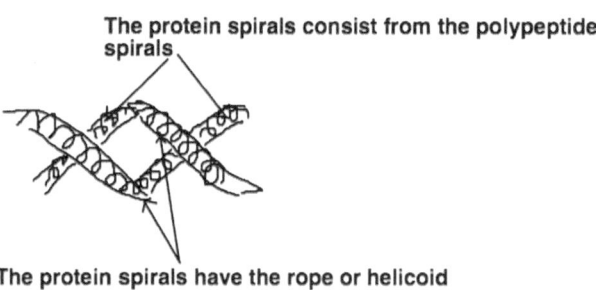

The protein spirals have the rope or helicoid structure

An archaic string with two archaic protein spirals. The archaic protein spiral has the archaic polypeptide spiral inside. Archaic polypeptide spirals consisted of polypeptides that attracted each other along their lengths.

Archaic protein spirals with opposite electrical charges created unstable forms of strings due to their metabolic needs; they assembled for short periods during metabolism. They had opposite metabolic needs due to different concentrations of the same coenzymes and metabolites that had made possible attraction and metabolism of metabolites.

The opposite pairs of the archaic strings moved in opposite directions producing different concentrations of the same metabolites. One archaic protein spiral had characteristics of archaic synthetic polypeptide spirals due to the attractions of more complex sugar molecules than another one. Another had characteristics of archaic decomposition protein spirals due to the attraction of simple metabolites of sugar molecules. They had the same function as synthetic and decomposition protein spirals.

Archaic protein spirals synthesized polypeptides, polysaccharides, fatty acids, fats, nucleotides, and other complex molecules from simple metabolites and decomposed the complex molecules into simple ones due to the movement and interaction of metabolites during metabolism.

Attraction and movement of opposite archaic protein spirals of archaic strings transformed pieces of nucleic acids (genes) of polypeptides into single nucleic acid spirals. The movements of these protein spirals by interaction among polypeptides blended nucleic acids (genes) of synthetic and decomposition polypeptides of protein spirals in pairs of nucleic acids with different qualities and quantities of nucleotides.

Pairs of archaic nucleic acids connected each other in single nucleic acid spirals. The same polypeptides of synthetic and decomposition protein spirals had different qualities and quantities of amino acids. Polypeptides of archaic protein spirals interacted and connected the pairs of genes also according to the order.

Archaic single nucleic acid spirals attracted complementary nucleotides to form ds nucleic acids. Archaic strings had two ds RNA between the protein spirals of strings. The strings of the unstable formation transformed into strings of a stable formation simultaneously so that single nucleic acid spirals were formed. Stable forms of archaic strings enabled a novel way to biosynthesize protein spirals and interchange parts of protein and nucleic acid spirals similar to contemporary strings. These archaic strings existed in pairs of strings like the pairs of filament strings.

Archaic strings did not undergo division by the chewing gum phenomenon. The long archaic strings had broken into short ones under unknown environmental conditions.

Association of archaic synthetic and decomposition strings produced an organization of filaments or complex cylinders under unknown environmental conditions. Filaments or complex cylinders had unity and the conflict of opposites. Filaments negated archaic strings and made the evolutionary jump to further development of prokaryotic cells. Archaic complex cylinders negated archaic strings and made the evolutionary jump to further development of archaic eukaryotic cells. I explain archaic complex cylinders in the hypothetical development of archaic and contemporary eukaryotic cells.

Archaic Filaments

Archaic synthetic and decomposition strings attracted and twisted around each other forming growing and nongrowing parts at their opposite parts. Filaments had the same structure as contemporary filaments have of prokaryotic cells, and they interchanged parts of polypeptide and nucleic acid spirals with filaments of the pairs. They had grown and had divided also as contemporary filaments do.

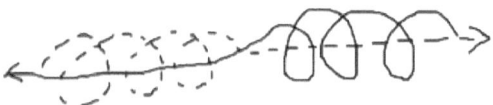

Model of contemporary filaments presents model of archaic filaments.

Archaic filaments had the winding phenomenon before division and the unwinding phenomenon after division. They lived in labile groups of separate pairs of filaments that had been poorly connected metabolically. The paired groups of archaic filaments had allowed them better metabolism, growth, and division due to an increased amount of metabolites and protection from the environment.

Archaic filaments had fewer groups of gene pairs that enabled better survival under environmental conditions by performing phenotype variations the same as contemporary prokaryotic cells do. Their genes had mutated under changing environmental conditions. Archaic mutated genes produced new proteins, which enabled them to survive under different environmental conditions.

Unknown environmental conditions caused evolution of labile groups of pairs of filaments into prokaryotic cells. These pairs with stronger opposite electromagnetic attraction attracted each other into stable groups of forms that were predecessors of archaic prokaryotic cells. Pairs of filaments with stronger opposite electromagnetic attraction in archaic prokaryotic cells negated the pairs of archaic filaments with weaker opposite electromagnetic attraction. They had united as opposites. Archaic prokaryotic cells made an evolutionary jump.

Archaic Prokaryotic Cells

Archaic prokaryotic cells had different forms and sizes likely due to unstable attractions of archaic filaments. Archaic prokaryotic cells had different forms and sizes likely due to a lack of peptidoglycan and polysaccharides molecules necessary for building cell walls, which had given stable forms for cocci, bacilli, spirals, and others. Peptidoglycan and polysaccharides had pressed archaic filaments into stable forms. Some archaic prokaryotic cells had polypeptides of polypeptide spirals of strings that could synthesize NAG and NAM molecules. They also produced polysaccharides, peptidoglycan, fats, and other complex molecules.

Different environmental conditions caused mutation of gene pairs at the cells and elimination of cells without walls. The mutated genes produced new protein spirals that enabled better attraction of the pairs of archaic filaments, better metabolism, and an evolutionary jump to contemporary prokaryotic cells.

Under the influence of the environment, prokaryotic cells transformed into stable forms because their filaments enabled them adequate metabolism according to environmental conditions. The cells developed a greater number of gene pairs during their evolution due to steady changes in their environmental conditions.

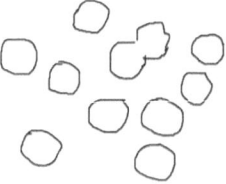

Methanococcus jannaschii are archaic prokaryotic cells that did not change or changed little over a long time.

Environmental conditions also forced stable forms of prokaryotic cells to evolve further; their filaments attracted each other and released complex carbohydrates and peptidoglycan transforming archaic prokaryotic cells into contemporary ones.

Contemporary Prokaryotic Cells

Contemporary prokaryotic cells started with the appearance of stable forms of prokaryotic cells that repeated after binary fission. These cells produced outside polysaccharides and peptidoglycan molecules that had covered and pressed their cytoplasm with filaments. The cell membranes consist of squeezed filaments due to the pressure of the cell wall; those filaments give the impression of a flat membrane in electron microscope images.

Polysaccharides and peptidoglycan molecules protected and prevented their filaments from attracting other filaments producing stable forms of prokaryotic cells and optimal conditions for life inside despite harsh environmental conditions. Contemporary prokaryotic cells undergo continuous changes due to gene mutation of filaments and natural selection. Genes mutated due to different environmental conditions. Evolution of prokaryotic cells never ends due to continuous changes in the environment.

The contemporary approach to the phylogenetic tree of forms of life does not explain the origin of prokaryotic and eukaryotic cells, their evolution, and their different characteristics. The new approach to the essential characteristics of prokaryotic and eukaryotic cells produces hypothetical but logical explanations of the evolution of their structures and archaic predecessors of prokaryotic and eukaryotic cells into stable ones. The new approach to the evolution of prokaryotic and eukaryotic cells will change according to new facts and new understanding.

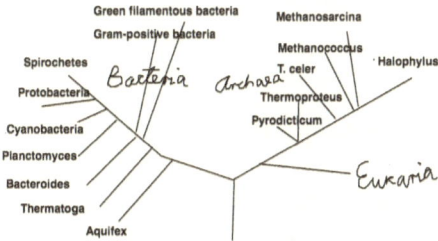

Domain bacteria and domain archaea include prokaryotic cells that lack membrane-enclosed nuclei and organelles, whereas domain eukarya include eukaryotes and more-complex organisms that contain membrane-bound nuclei and organelles according to the literature.

Universal Laws of Nature and Prokaryotic Cells

The laws of dialectic materialism are philosophical descriptions of universal laws of nature (appendix I) that explain also characteristics of prokaryotic cells. The laws of dialectic materialism as universal laws of nature contain and unite all other known and unknown laws of nature that also regulate different characteristics of matter. Dialectical laws work together and cannot separate from each other just as matter and energy cannot.

Universal laws of nature describe the essential characteristics of the cosmos. Spiral galaxies in collision, Image Credit: Debra Meloy

Elmegreen, Vassar College, et al., and the Hubble Heritage Team. (AURA/STScI/NASA).

The dialectic laws are the law of unity and conflict of opposites, the law of the passage of quantitative changes into qualitative changes, and the law of the negation of the negation.

The Law of Unity and Conflict of Opposites

Filaments attract each other by building different forms and sizes of prokaryotic cells according to the law of unity and conflict of opposites. Filaments consist of two strings, decomposition and synthetic strings, that interchange metabolites in the zigzag model. The straight part of a synthetic string combines with the coil part of the decomposition string and vice versa. The strings are unified as they conflict.

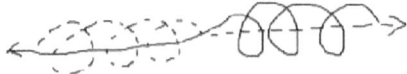

Decomposition strings produce decomposition protein spirals in synthetic protein spirals that move from their straight parts to their coil parts. Decomposition strings have decomposition protein spirals with tRNA that move from coil parts to straight parts while synthetic strings produce synthetic protein spirals in decomposition protein spirals that move from its straight parts to its coil parts.

Synthetic strings have synthetic protein spirals with tRNA that moves from the coil parts to the straight parts, the opposite of decomposition protein spirals. Polypeptides of different protein spirals that have opposite movements unify strings and the conflict of opposites by their different concentrations of the same coenzymes, metabolites, and electrical charges.

Strings make the unity of filaments and conflict of opposites by their specific qualities and organization in filaments and the opposite movements by their opposite electromagnetic attractions. Filaments of pairs unify the cell by their attraction among themselves and conflict of opposites by their opposite electrical charges and opposite movements.

The different polypeptides of decomposition and synthetic protein spirals of the different strings of the filaments assembling and disassembling NAG and NAM, peptidoglycan, and metabolites also constitute the unity and conflict of opposites.

Filaments with characteristics opposite those of strings, strings with characteristics opposite those of protein spirals, and protein spirals with characteristics opposite those of polypeptides enable the opposite movement of the structures by each other, metabolism, growth, and binary cell fission. The opposite structures of filaments with opposite movements make the unity and conflict of opposites simultaneously at prokaryotic cells.

The Law of the Passage of Quantitative into Qualitative Changes

Prokaryotic cells consist of pairs of filaments with different qualities, lengths, and numbers. Polypeptides with the same coenzymes of different concentrations at different filaments determine different concentrations of metabolites and characteristics of prokaryotic cells according to the law of the passage of quantitative into qualitative changes.

The different polypeptide spirals of decomposition and synthetic protein spirals have different specific polypeptides with different lengths and qualities. These differences attract different concentrations of coenzymes, which in turn attract different concentrations of the same and different metabolites.

Differing lengths and qualities of polypeptides and differing lengths and numbers of the pairs of the filaments determine the characteristics of prokaryotic cells. These include their metabolism, qualitative and quantitative biochemical contents, cell walls, aerobic and anaerobic qualities, ways of oxidation of H+ ions, survival rates under different salinities, temperatures and pH, utilization of energy, production of molecules, frequency of depolarization and polarization, bioelectricity, and other unknown characteristics.

The Law of the Negation of the Negation

The first forms of life transformed through evolution prompted by differing environmental conditions that acted and will act as the law of the negation of the negation. The building of hypothetical archaic forms of prokaryotic cells was due to the law of the negation of the negation.

Specific environmental conditions forced archaic polypeptides with coenzymes and their metabolites to attract each other along their lengths into polypeptide spiral structures according to their attraction affinities. The polypeptide spirals transformed into protein spirals by attractions of the coils of the polypeptide spirals in themselves. Environmental conditions negated archaic polypeptides and transformed them into protein spirals.

Specific environmental conditions forced archaic protein spirals to attract each other along their lengths into pairs in helicoid structures according to their attraction affinities. The archaic polypeptides of the coils of the two archaic protein spirals had opposite concentrations of the same coenzymes and ions, which enabled their attraction into archaic strings, movement, and metabolism. Archaic strings had unity and the conflict of opposites; they had novel biosynthesis of protein and nucleic spirals and interchange of parts of protein and nucleic acid spirals among the pairs of strings. Specific environmental conditions negated protein spirals and transformed them into string pairs.

Environmental conditions forced pairs of archaic strings to attract each other along their lengths into pairs of filament structures according to their attraction affinities and transformed them into pairs of archaic filaments, which showed unity and conflict of opposites by synthetic and decomposition archaic strings that had twisted around each other forming growing and nongrowing parts at opposite parts and thus making the unity and conflict of opposites.

Archaic filaments had the same structure as contemporary filaments do of prokaryotic cells and interchanged parts of archaic polypeptide and nucleic acid spirals among the pair of the archaic filaments. They grew and divided as contemporary filaments do. Unknowns in the environment negated the pairs of the archaic strings and transformed them into pairs of archaic filaments.

Unknown environmental conditions caused evolution of the labile groups of the pairs of filaments into prokaryotic cells. The pairs with stronger opposite electromagnetic attractions attracted each other into stable groups of forms that had been archaic prokaryotic cells. Unknown environmental conditions negated the pairs of archaic filaments with weaker opposite electromagnetic attractions and transformed them into archaic filaments with stronger opposite electromagnetic attractions of archaic prokaryotic cells.

Unknown environmental conditions had also caused evolution of archaic prokaryotic cells into contemporary cells. Due to environmental conditions, archaic prokaryotic cells transformed into more-stable forms of contemporary prokaryotic cells that had different genetic groups.

Different genetic groups of filaments of prokaryotic cells had the ability by phenotype variation to adjust to environmental conditions and had fewer groups of gene pairs than did archaic prokaryotic cells. The gene pairs evolved due to steady changing of environmental conditions, which forced archaic prokaryotic cells to evolve so their filaments attracted each other more and they released complex carbohydrates and peptidoglycan. The environment negated archaic prokaryotic and transformed them into contemporary prokaryotic cells.

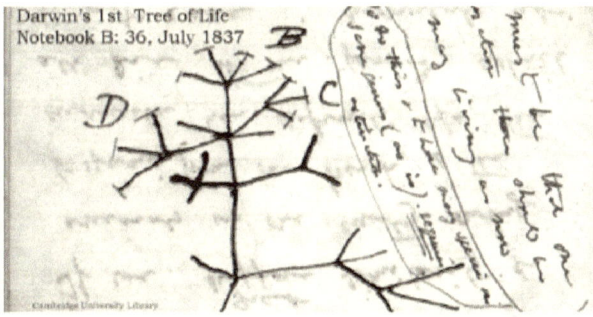

Darwin's first tree of life is based on the law of the negation of the negation whereby archaic animals or plants negated previous species, which negated their predecessors to survive in cruel environmental conditions. Environmental conditions will always effect change in species by mutation.

CHAPTER 15
PRACTICAL TAXONOMY OF PROKARYOTIC CELLS

Contemporary taxonomy of prokaryotic cells is not practical for investigation of a huge number of different prokaryotic cells and ecosystems due to the chaotic, complex, and unclear definitions of prokaryotic species and strains (appendix XXXIX).

Contemporary taxonomy is replaced with practical taxonomy of prokaryotic cells according to the universal laws of nature. It is designed to explore a huge number of prokaryotic cells and their ecosystems for industrial and medical microbiology needs.

Taxonomy of prokaryotic cells relies on the new view of prokaryotic cells' structures and the universal laws of nature that also determine different characteristics of prokaryotic cells. The methods are explained in appendix XLI. Practical taxonomy has these tasks:

- to simplify investigation of a huge number of different prokaryotic cells of different ecosystems
- to explain different characteristics of prokaryotic cells according to the results of the new methods
- to explain why different prokaryotic cells live in environments with different characteristics (air, without air, different pH, high salt concentrations, the sun, and high and low temperatures)
- to explain why prokaryotic cells live in different prokaryotic ecosystems regarding their metabolic connections
- to store different prokaryotic cells in a logical way for practical needs of industrial and medical microbiology
- to help create artificial prokaryotic ecosystems for industrial, agricultural, and medicinal needs

Industrial microbiology investigates bacterial ecosystems and their species in different environments to

- find the best bacterial ecosystems for utilization of waste,
- enrich ground with artificial prokaryotic ecosystems to improve agriculture,
- enrich animal food with artificial prokaryotic systems for better growth and disease prevention, and
- produce organic compounds by prokaryotic cells for industrial, agricultural, medical, and other needs.

Medical, veterinary, and plant microbiology investigates prokaryotic ecosystems to

- find pathogenic bacteria and bacterial ecosystems that cause infectious diseases in humans, animals, and plants;
- design prokaryotic ecosystems that eliminate pathogens and other highly resistant prokaryotic cells in humans and animals (such ecosystems will be used for colonization of the skin and guts of humans and animals and their environments); and
- promote better digestion and nutrition of humans and animals by mixing prokaryotic ecosystems mixed with food.

A huge number of prokaryotic cells are not investigated and classified according to the contemporary taxonomy due to different reasons:

- Simple definitions of prokaryotic species and strains do not exist (appendix XXXIX).
- Different and expensive taxonomic methods are not used uniformly.
- Many prokaryotic cells do not grow in artificial media.

The taxonomic names of prokaryotic species are also not connected with their characteristics for the practical needs of industrial and medical microbiology. There are many examples: *Pseudomonas aeruginosa, Staphylococcus aureus, Salmonella spp.,* and many others do not explain their essential characteristics. The groups of the existing classification (domains, phyla, classes, subclasses, orders, suborders, families, and genera) are not connected with their metabolic and other characteristics. The existing classification is good only for a scholastic approach and limited practicality.

Prokaryotic ecosystems and cells cannot be taxonomically investigated and used because simple definitions of prokaryotic species based on their shapes, walls, environmental conditions, metabolic characteristics, enzymes, and genes do not exist; names of prokaryotic species and their strains do not have general descriptions; and prokaryotic cells do not grow in natural and standard media for investigation.

A more-practical and simpler system of taxonomy will help microbiologists investigate prokaryotic ecosystems and cells based on different needs and simple grouping based on shape, cell walls, survival rates under standardized environmental conditions, metabolic characteristics, and universal laws of nature. New names for prokaryotic species will include their general characteristics to simplify their storage and use. Universal laws of nature associated with this new approach to prokaryotic cells will make their taxonomy practical.

The law of unity and conflict of opposites helps us understand connections between metabolism and the structures of prokaryotic cells built by polypeptides that build protein spirals, strings, and filaments. Protein spirals move in opposite directions during metabolism as the conflict of opposites makes the unity of the strings.

The law of quantitative changes becoming qualitative ones helps us understand how different concentrations of coenzymes connected with polypeptides determine their characteristics. Different concentrations depend on the lengths of polypeptides and protein spirals in strings and filaments. The law of the negation of the negation helps us understand the evolution of archaic prokaryotic cells due to harsh environments.

In summary, practical taxonomy of prokaryotic species negates their contemporary classification due to the industrial and medical microbiology need to solve many problems (handling waste, improving agriculture, developing resistant pathogenic bacteria and many others).

Specific Characteristics of Prokaryotic Species Based on Practical Taxonomy

The characteristics of prokaryotic cells according to the strict order are used to develop a practical taxonomy and the new methods explained in appendix XLI. The specific characteristics of prokaryotic cells are next.

1. Shape of Cells and Presence of Spores

Prokaryotic cells' shapes depend on the filaments, which have different lengths and qualities. The shapes of different prokaryotic cells are these:

1. Coccus
2. Rod or bacillus
3. Diplococci (cocci in pairs)
4. Neisseria (coffee-bean shape in pairs)
5. Coccobacilli
6. Tetrads (cocci in packets of four)
7. Sarcinae (cocci in packets of eight, sixteen and thirty-two cells)
8. Mycobacteria
9. Corynebacteria (Chinese letters arrangements)
10. Streptococci (cocci in chains)
11. Micrococci or staphylococci
12. Spore-forming rods
13. Streptomycetes (mold like filamentous bacteria)
14. Vibrios
15. Spirochetes
16. Spirilia
17. Amorphous

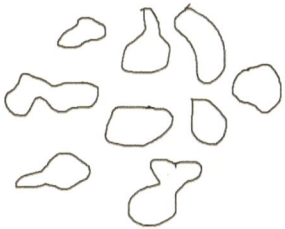

Mycoplasma species have amorphous shapes.

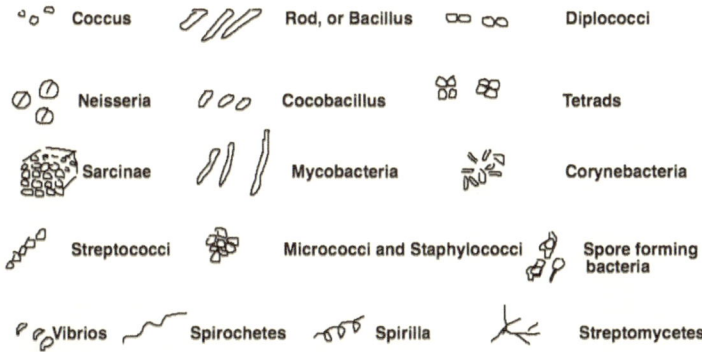

Different shapes of prokaryotic cells are due to different sizes and qualities of filaments.

2. Cell Walls

Cell walls depend on abilities of their plasma membranes to attract parts of protein with its fats and nucleic acid spirals. Prokaryotic cells can be gram-positive, gram-negative, gram-variable, or not gram stained.

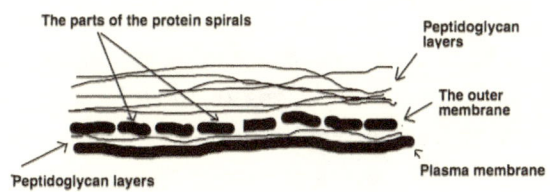

The cell wall of Gram-negative prokaryotic cells

Cell walls of gram-negative bacteria have outer membranes.

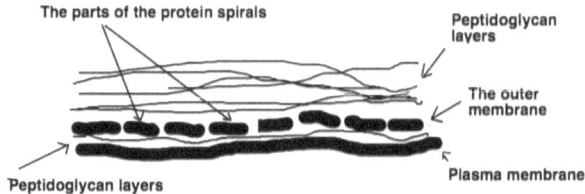

The cell wall of Gram-negative prokaryotic cells

Cell wall of gram-positive bacteria do not have outer membranes.

3. Survival of Species with and without Air, High and Low Ph, High Salt Concentration, Sun, Darkness, and High and Low Temperatures

Aerobic and anaerobic survival depends on concentrations of the porphyrin-Fe++ complexes, which appear in groups along the strings. Prokaryotic cells that have the smallest concentrations of the porphyrin-Fe++ complexes are anaerobic while those with the largest concentrations are aerobic. The facultative prokaryotic cells have concentrations of porphyrin-Fe++ complexes between the extremes.

Aerobic prokaryotic cells oxidize hydrogen atoms only by oxygen atoms. Anaerobic prokaryotic cells oxidize hydrogen atoms only by negative metabolites from the Krebs cycle. Facultative prokaryotic cells can oxidize hydrogen atoms by oxygen atoms under aerobic conditions or the negative metabolites from Krebs cycle under anaerobic ones. The way of oxidation of hydrogen ions tells us how prokaryotic cells oxidize hydrogen ions and where they live. There are three groups according to their surviving under air or without air marked with the capital letters. The three groups can further group according to their surviving under different conditions of pH of water, salinity, Sun, and temperature present in environments. They can survive under different environmental characteristics only if they have inside specific characteristics. Other words, their surviving under specific different specific characteristics discover their inside specific characteristics.

Aerobic prokaryotic cells	Facultative prokaryotic cells	Anaerobic prokaryotic cells
A	F	W

Growth of prokaryotic cells under different pH

Low pH (5)	Low pH (5)	Low pH (5)
no growth (0)	small growth (s)	normal growth (G)

High pH (8)	High pH (8)	High pH (8)
no growth (0)	small growth (s)	normal growth (G)

Growth of grouped prokaryotic cells under high salt concentrations

High salt concentration (60%)	High salt concentration (60%)	High salt concentration (60%)
no growth (0)	small growth (s)	normal growth (G)

Growth of prokaryotic cells under the Sun or darkness

Sun	Sun	Sun
no growth (0)	small growth (s)	normal growth (N)

Growth of prokaryotic cells under low and high temperatures

Low temperature (10C)	Low temperature (10C)	Low temperature (10C)
no growth (0)	small growth (s)	normal growth (G)

Medium temperature (25C)	Medium temperature (25C)	Medium temperature (25C)
no growth (0)	small growth (s)	normal growth (G)

High temperature (50C)	High temperature (50C)	High temperature (50C)
no growth (0)	small growth (s)	normal growth (G)

4. Speed of Metabolism

Speed of metabolism determines speed of growth of prokaryotic cells in standard media of standard volume and temperature in a unite of time. Speed of growth of prokaryotic cells in media (S), growth of prokaryotic cells (G) and time (t) have relationships which can be expressed this formula.

$$S = G/t$$

Growth of prokaryotic cells in media manifests as increase of optic densities of media. Therefore, speed of growth is calculated according to this formula.

$$S = (OD1-OD0) / (t1-t0)$$

S = Speed of growth of a prokaryotic species in a standard medium and volume

OD0 = Initial optic density of standard media with added prokaryotic cells without their growths

t0 = Initial time of the addition of standard concentrations of prokaryotic cells without their growth (OD0)

OD1 = Optic density of standard media with added the standard concentration of prokaryotic cells with their growths

t1= Incubation time of added prokaryotic cells with their growth (OD1)

Speed of metabolism or growth of prokaryotic cells depends on coenzymes and metabolites concentrations in the protein spirals of prokaryotic cells. Prokaryotic cells with a large quantity of coenzymes and their metabolites in the protein spirals metabolize and grow quickly; the reverse is the case for cells with smaller quantities of coenzymes and their metabolites. There are three groups of prokaryotic cells with different powers of metabolisms marked with capital letters.

High power of metabolism	Medium power of metabolism	Low power of metabolism
H	M	L

Proteus spp. with their speed of metabolism is a measure of high speed of metabolism (H). *Streptococcus spp.* with their speed of metabolism are a measure of medium speed of metabolism (M). *Mycobacterium spp.* with their speed of metabolism are a measure of the low speed of metabolism (L). The speed of metabolism of different prokaryotic spp. will be determined by their speed of metabolism closest to one of the three speeds of metabolism.

The speed of metabolism for prokaryotic cells of the same species changes if the standard medium is changed in terms of content, volume, or temperature.

The different growths of metabolism of prokaryotic species of same species under influence of the coenzymes and the metabolites of the standard concentrations in the standard medium are used for finding indirectly concentrations of specific coenzymes and their metabolites and quantitative relationships among the coenzymes and their metabolites. The quantitative relationships among the coenzymes and the metabolites of specific prokaryotic species determine their metabolic and other characteristics. Information of metabolic and other characteristics of different prokaryotic species are necessary for investigation of different prokaryotic ecosystems and creation of different prokaryotic systems for different needs.

Determining Quantitative Concentration of Coenzymes and Their Metabolites

The phenomenon of the different speed of metabolism or growth of prokaryotic cells of the same prokaryotic species is caused by separately adding coenzymes or metabolites to standard concentrations, media, volumes, and temperature with standard concentrations of prokaryotic cells of the same species. The phenomenon is explained by the changing of speed of metabolism of the species under the influence of different concentrations of coenzymes or metabolites.

The energy of the coenzymes or metabolites with their Brownian motion in standard media of standard volume influences the energy of the coenzymes attracted by polypeptides of protein spirals of prokaryotic cells or the energy of metabolites attracted by coenzymes of prokaryotic cells. Coenzymes or metabolites added separately to standard media with prokaryotic cells of the same species act with their own energy on that of their specific coenzymes and their specific metabolites in two hypothetical ways—decreased or increased metabolism.

Decreased Metabolism in Prokaryotic Cells

Decreased metabolism of prokaryotic cells occurs if their polypeptides attract less gmol/l concentration of the specific coenzymes than gmol/l concentration of the free specific coenzymes in the medium. Energy of coenzymes with more gmol/l concentration in media moves to the attracted coenzymes with less gmol/l concentration and energy. This movement increases the energy of the attracted coenzymes. Polypeptides with the same power attract fewer coenzymes due to their increased energy; the consequence is decreased metabolism.

Decreased metabolism occurs also if their coenzymes attract less gmol/l concentration of the specific metabolites than gmol/l concentration of the free specific metabolites in the medium. Coenzymes with the same attraction power attract fewer specific metabolites due to their increased energy; the consequence is decreased metabolism.

Increased Metabolism

Increased metabolism occurs when polypeptides attract more gmol/l concentration of the specific coenzymes than gmol/l concentration of the free specific coenzymes in the medium. The consequence is that the energy of attracted coenzymes with more gmol/l concentration moves to the free specific coenzymes with less gmol/l concentration and energy; the movement decreases the energy of the attracted coenzymes. Polypeptides with the same attraction power attract more coenzymes due to their decreased energy; that increases metabolism.

Increased metabolism in prokaryotic cells takes place also if their coenzymes attract more gmol/l concentration of the specific metabolites than gmol/l concentration of the free specific metabolites in the medium. Energy thus moves to metabolites with less gmol/l concentration and energy. Coenzymes with the same attraction power attract more metabolites due to their increased energy, and that increases metabolism.

Speed of metabolism or growth is directly proportional to concentrations of coenzymes and metabolites in protein spirals of filaments and indirectly proportional to concentrations of coenzymes and metabolites in media. Speed of metabolism is indirectly proportional to concentrations of coenzymes and metabolites in media because they influence with their energy on concentrations of coenzymes and metabolites attracted with protein spirals of prokaryotic cells. The relation of speeds of metabolism among specific coenzymes concentrations in protein spirals of prokaryotic cells and specific coenzyme concentration is expressed by this formula.

$S = k\, c/C$

S = speed of metabolism or growth of prokaryotic cells

c = concentrations of specific coenzymes in the protein spirals of prokaryotic cells

C = concentrations of specific coenzymes in media

k = corrective factor enables the equation (it is found by experiments)

Different speeds of metabolism of prokaryotic cells under the influence of different compounds can be used to calculate gmol/l concentration of the attracted coenzymes and the metabolites of prokaryotic cells in the standard medium.

Calculation of gmol/l Concentration of Coenzymes or Metabolites of Prokaryotic species in Standard Media

Concentration of specific coenzymes in a unite of time can be calculated according to the formula which ensued from the above formula ($S = kc/C$).

c = SC/ k

c = concentration of specific coenzymes in gmol/l in the protein spirals of prokaryotic cells in a unite of time

S = speed of growth of prokaryotic species

C = concentration of specific coenzymes in gmol/l in media

k = corrective factor

Concentrations of specific coenzymes in the standard media are unknown. They can be calculated according to the formula:

C = S a/ (S' − S)

S = speed of growth of prokaryotic species in standard media without added coenzymes

C = unknown concentration of coenzymes

S' = speed of growth of prokaryotic species in standard media with added coenzymes

a = added known concentration of specific coenzymes

The formula ensues from reasoning that speeds of growth of same prokaryotic species in the standard medium with unknown concentration of coenzymes without and with added concentration of coenzymes have mathematical relation with unknown concentration of coenzymes and added concentration of coenzymes according to this formula.

S : C = S' : (C + a)

S (C + a) = C S'
S C + S a = C S'
S a = C S' − SC
S a = C (S' − S)
C = S a/ (S' − S)

5. Quantitative Relationships among Metabolically Connected Coenzymes in Prokaryotic Cells

The results of the calculations of the unknown concentrations of the coenzymes attracted by prokaryotic cells' protein are put in this table.

ATP	B_1	Folic acid	Biotin	CoA	K	B_6	NAD	FMN

The concentrations of the coenzymes and the metabolites in gmol/l determine the quantitative relationships among the coenzymes, and that determines if the prokaryotic species belong to the decomposition or the synthetic group. That also determines if prokaryotic cells have low or high frequency of depolarization. Measurement of extremely small concentration of coenzymes in the protein spiral of prokaryotic cells is very difficult. Therefore, those concentrations are determined by the speed of growth of prokaryotic cells under the influence of coenzymes of known concentrations in gmol/l and also expressed in gmol/l. The experiments will prove or disprove this approach.

6. Synthetic and Decomposition Groups

Prokaryotic cells synthesize and decompose simultaneously with prevailing decomposition or synthesis. The quantitative relationships among metabolically connected coenzymes are established by quantitative relationships among the speeds of metabolism under the influence of coenzymes metabolically connected.

If the prokaryotic cells have the below quantitative relationships among their coenzymes, they are in the synthetic group.

ATP > B_1 > Folic acid > Biotin and Folic acid, CoA < K < ATP

If the prokaryotic cells have the below quantitative relationships among their coenzymes, they are in the decomposition group.

ATP < B_1 < Folic acid < Biotin and Folic acid, CoA > K > ATP

If prokaryotic cells have a balance between synthesis and decomposition with slight prevailing synthesis or decomposition, they are the medium ones. In summary, there are three groups that are also marked with the capital letters.

Synthetic prokaryotic cells	Synthetic / Decomposition prokaryotic cells	Decomposition prokaryotic cells
S	B (both)	D

7. Frequency of Depolarization

NAD coenzymes attract hydrogen atoms from metabolites as positive ions (H+) and transfer them to FMN coenzymes. When FMN coenzymes are saturated with H+, they release them because they do not have enough energy to attract them. Released hydrogen ions influence other metabolites attracted by other coenzymes, which release metabolites in a domino effect causing depolarization. The frequency of depolarization determines a quality of the released metabolites in the media and builds complex metabolites. Prokaryotic cells that depolarize frequently produce less complex metabolites and fewer simple metabolites because they have less time to biosynthesize complex ones. Prokaryotic cells that depolarize less frequently produce more complex metabolites and fewer simple metabolites because they have more time for biosynthesis of complex ones.

Frequency of depolarization is determined by the speed of metabolisms under influence of NAD and FMN coenzymes in standard concentration in standard media, calculations of their concentration, and the reasoning. This approach is due to a lack of technical devices to measure the frequency and energy of depolarization in prokaryotic cells. The frequency of depolarization in prokaryotic cells depends on concentrations of NAD and FMN coenzymes compared with concentrations of other coenzymes. NAD and FMN coenzymes with greater concentrations than other coenzymes have release H+ ions faster causing high frequency of depolarization. The reverse is true for those coenzymes with less concentrations.

Three groups of prokaryotic cells with different frequencies of depolarization are marked with capital letters.

High frequency of depolarization	Medium frequency of depolarization	Low frequency of depolarization
H	M	L

Prokaryotic cells with higher or lower frequencies of depolarization can have bigger NAD coenzymes concentrations than FMN ones or bigger FMN coenzymes concentrations than NAD ones. If the NAD concentrations are bigger than FMN (NAD > FMN), NAD coenzymes saturate FMN coenzymes with H+ ions faster. As a consequence, FMN coenzymes release them faster causing a higher frequency of depolarization. If the NAD concentrations are less than FMN (NAD < FMN), NAD coenzymes saturate FMN with H+ ions at a slower rate. FMN coenzymes release them slower causing a lower frequency of depolarization.

8. Enzymes and Their Quantities

Different prokaryotic cells utilize or do not utilize different complex molecules. The table below contains the different complex molecules and fictitious examples of the present (1) and the lacking enzymes (0) for their utilization at the same prokaryotic cells.

Lact-ase	Galact-ase	Mann-ase	Saccha-rase	Lypase	Cellulase	Starch enzyme	PAH*-ase	Urease	PET, MEHT**-ase	Collage-nase	DNAase
1	0	0	1	0	0	1	1	1	1	0	0

*Polycyclic aromatic hydrocarbons (PAHs) are mutagenic, cytotoxic, and carcinogenic organic chemicals widely distributed in the environment due to incomplete combustion of organic matter, emissions, automobile exhaust, domestic matter, and other factors. As an example of enzymatic bioremediation, PAH detoxification can be achieved by the use of laccases (enzymes capable of catalyzing the oxidation of phenols, polyphenols, and anilines, coupled to the 4-electron reduction of molecular oxygen to water).[248]

**PET stands for polyethylene terephthalate, a plastic with good mechanical, barrier, and optical properties. The results strongly suggest that ISF6_0224 protein is responsible for the conversion of mono (2-hydroxyethyl) terephthalic acid (MEHT) into PET's two environmentally benign monomers, terephthalic acid and ethylene glycol. As such, the team decided that ISF6_0224 should be termed a MEHT hydrolase abbreviated to MEHTase. [119]

Prokaryotic cells of the same species have the same enzymes that utilize different complex molecules. The strains of the same prokaryotic species have the same enzymes of different quantities. The same strains of the same prokaryotic species have the same enzymes of the same quantities. The quantities of the specific enzymes of the same prokaryotic cells can be measured by the power of metabolism for a specific time produced by enzymes that act on complex molecules of the standard concentration in standard media.

9. Quantities of Genes and Names according to Contemporary Taxonomy of Prokaryotic Cells

Prokaryotic cells have different quantities of genes that produce different quantities of specific polypeptides. The contemporary literature has very limited information regarding analysis of different genes at different prokaryotic species. This characteristic will be explored in the future. Many prokaryotic cells have names according to Contemporary taxonomy. Huge number of prokaryotic cells are not examined and classified and therefore they do not have names. Finding name of examined prokaryotic species according to Contemporary taxonomy is important due to correlation of their characteristics with characteristics of known prokaryotic cells according to the literature that deals with prokaryotic cells classified according to Contemporary taxonomy.

Summarized Characteristics of Prokaryotic Species

The characteristics of prokaryotic species which are examined according to the new method (appendix XLI) are summarized in the nine groups in the below table.

1.Shape	2.Cell wall	3.Environ.	4.Speed of metabolism	5. Conc. of coenzymes	6.Synth. or Decom.	7.Freq. of depolarize.	8.Enzymes	9.Name

3. Environmental characteristics

Oxidation	pH 5	pH 8	High salt concentration	Sun's rays	10C, 30C	50C

5. Concentrations of coenzymes and metabolites

ATP	B$_1$	Folic acid	Biotin	CoA	B$_6$	K vitamins	NAD	FMN

8. Enzymes

Lactase	Gala-ctase	Mann-ase	Sacchar-ase	Lipase	Cellul-ase	Starch enzyme	PAH enzyme	Urease	PET, MEHT enzyme	Colla-genase	DNA-ase

Prokaryotic Species

Prokaryotic cells belong to a species according to the practical classification if they have:

1. same shape and presence or absence of spores;
2. same cell walls;
3. same survival with and without air, high salt concentration, Sun, darkness and similar survival under high and low pH and temperatures;
4. similar speeds of growth in standard media with small variations (+/-10%);
5. same quantitative relationships among metabolically connected coenzymes in prokaryotic cells;
6. belonging to same synthetic or decomposition group;

7. belonging to same high-or low-frequency depolarization group;
8. same enzymes of similar concentrations; and
9. same genes of the similar concentrations (this is not discussed).

Prokaryotic Strains

Prokaryotic cells of a species belong to a strain if they have the same characteristics of the prokaryotic species and have

1. same survival under high and low pH and temperatures,
2. very similar speed of growth, very similar concentrations of coenzymes attracted by prokaryotic proteins, and
3. same enzymes of very similar concentration.

Names of Prokaryotic Species and Their Strains

Modern information technology (IT) enables the complex names of prokaryotic species and their strains. The name of the prokaryotic species of the practical classification will have in itself the major characteristics of the prokaryotic species that are described by pictures, letters, and numbers. The major characteristics of prokaryotic species described by pictures, letters, and numbers are explained in the table as an example.

1 cocci	2 A	3 0	4 0	5 s	6 G	7 0	8 s	9 0.1/h	10 B	11 M	12 "name"

1. picture of the shape of prokaryotic species (cocci, bacilli, streptococci, spirals, or others) and spore;
2. prokaryotic species that belong to facultative, aerobic, and anaerobic group (A, F, or W);
3. prokaryotic species that grow under low pH (0, s, or G);
4. prokaryotic species that grow under high pH (0, s, or G);
5. prokaryotic species that grow under high salt concentrations (0, s, or G);
6. prokaryotic species that grows under the Sun (0, s or G);

7. prokaryotic species that grow under high temperatures (0, s, or G);

8. prokaryotic species that grow under low temperatures (0, s, or G);

9. Speed of growth of prokaryotic species expressed by a number;

10. Prokaryotic species belong to synthetic or decomposition group (S, B*, or D); (B* means between [B] the synthetic [S] and the decomposition group [D]);

11. Prokaryotic species belong to high-or low-frequency group (H, M**, or L); (M** means between the high frequency [H] and the low frequency [L])

12. Name of prokaryotic species according to Contemporary classification of prokaryotic cells.

Prokaryotic strains of the species have the name of the prokaryotic species and concentrations of coenzymes in gmol/l.

ATP	B_1	Folic acid	Biotin	CoA	B_6	K vitamins	NAD	FMN
x1	x2	x3	x4	x5	x6	x7	x8	9x

Information technology will allow easy communication among microbiologists regarding the names of prokaryotic species and strains according to the new taxonomy. A computer program can be created to translate pictures, letters, and numbers in the major characteristics of prokaryotic cells.

Classification of Prokaryotic Species according to Practical Taxonomy

The prokaryotic species and their strains with their new names are classified in the groups according to an order of the characteristics for collecting and storing purposes due to the usage of prokaryotic cells in industrial and medical microbiology.

The table below shows grouping of prokaryotic species according to the growth characteristics in different environmental conditions and the major characteristics.

Oxidation 1.	Low pH 2.	High pH 3.	High salt conc. 4.	Sun 5.	Low temp. 6.	High temp. 7.	Speed of metabolism 8.	Synthetic or Decomposition Metabolism 9.	High or Low freq. 10.
A, F, W	0, s, G	0, s, G	0, s, G	0, s, G	0, s, G	0, s, G	H, M, L	S, B, D	H, M, L
1-3	1-9	1-27	1-81	1-243	1-729	1-2187	1-6561	1-19683	1-59049

Prokaryotic cells are grouped in ten groups according to the strict order of the characteristics of the table. Each group has three subgroups. The first group of prokaryotic cells is according to the growth with and without air and has three subgroups.

Aerobic prokaryotic cells (A)	Facultative prokaryotic cells (F)	Anaerobic prokaryotic cells (W)

The second group is according to the growth under low pH and has three subgroups.

no growth (0)	small growth (s)	normal growth (G)

The third group is according to the growths under high pH and has three subgroups.

no growth (0)	small growth (s)	normal growth (G)

The fourth group is according to the growths under high salt concentrations and has three subgroups.

no growth (0)	small growth (s)	normal growth (G)

The first group has now 81 subgroups.

The fifth group is according to the growth under the Sun and has three subgroups.

no growth (0)	small growth (s)	normal growth (G)

The first group has now 243 subgroups.

The sixth group is according to the growth under low temperature and has three subgroups.

no growth (0)	small growth (s)	normal growth (G)

The first group has now 729 subgroups.

The seventh group is according to the growth under high temperature and has three subgroups.

no growth (0)	small growth (s)	normal growth (G)

The first group has now 2,187 subgroups

The eighth group of cells is according to the speed of metabolism and has three subgroups that are marked with letters.

High speed of metabolism (H)	Medium speed of metabolism (M)	Low speed of metabolism (L)

The first group has now 6,562 subgroups.

The ninth group of prokaryotic cells is according to if they are synthetic, decomposition, or both and has three subgroups.

Synthetic (S)	Between two extremes (B)	Decomposition (D)

The first group has now 19,683 subgroups (6,561 x 3).

The tenth group is according to the frequency of depolarization and has three subgroups.

Low frequency of depolarization (L)	Medium frequency of depolarization (M)	High frequency of depolarization (H)

Thus, the first group has 59,049 subgroups (19,683 x 3).

The number of prokaryotic cells varies between several millions and hundreds of millions according to the literature. They can be classified in 59,049 subgroups that can be classified in three major groups according to their growth with or without air.

- I-19,683 belong to the aerobic group;
- II-19,683 belong to the facultative group;
- III-19,683 belong to the anaerobic group.

Each of these groups has three subgroups of 6,561 according to the three different growths under the low pH.[5]

- A1-6,561 belong to the no-growth group (0) of the aerobic, facultative, or anaerobic group;
- B1-6,561 belong to the small-growth group (s) of the aerobic, facultative, or anaerobic group;
- C1-6,561 belong to the normal-growth group (G) of the aerobic, facultative or anaerobic group.

Each subgroup above has 6,561 subgroups according to the three different growths under the high pH.

- A2-2,187 belong to the no-growth group (0) of A1, B1, or C1;
- B2-2,187 belong to the small-growth group (s) of A1, B1, or C1;
- C2-2,187 belong to the normal-growth group (G) of A1, B1, or C1.

Each of these has three subgroups according to salt concentration.

- D-729 belong to the no-growth group (0) of A2, B2, or C2;
- E-729 belong to the small-growth group (s) of A2, B2, or C2;
- F-729 belong to the normal-growth group (G) of A2, B2, or C2.

Each of these has three subgroups according to the energy they derive from the Sun.

- G-243 belong to the no-growth group (0) of D, E, or F;
- H-243 belong to the small-growth group (s) of D, E, or F;
- I-243 belong to the normal-growth group (G) of D, E, or F.

Each of these has three subgroups based on growth under low temperatures (9C).

- J1-81 belong to the no-growth group (0) of G, H, or I;
- K1-81 belong to the small-growth group (s) of G, H, or I;
- L1-81 belong to the normal-growth group (G) of G, H, or I.

Each of these has 27 subgroups based on growth under high temperatures (50C).

- J2-27 belong to the no-growth group (0) of J1, K1, or L1;
- K2-27 belong to the small-growth group (s) of J1, K1, or L1;
- L2-27 belong to the normal-growth group (G) of J1, K1, or L1.

Each of these has 3 subgroups based on power of metabolism.

- a-9 belong to the group with high power of metabolism of J2, K2, or L2;
- b-9 belong to the group with medium power of metabolism of J2, K2, or L2;
- c-9 belong to the group with lower power of metabolism of J2, K2, or L2.

Each of these has three subgroups according to the three types of metabolism.

- a1-3 belong to the group of synthetic metabolism of a, b, or c;
- b1-3 belong to the group of synthetic and decomposition metabolism of a, b, and c;
- c1-3 belong to the group of the decomposition metabolism of a, b, or c.

Each of these groups of prokaryotic species has three subgroups according to frequency of depolarization.

- a2-1 belongs to the group with high frequency of depolarization of a1, b1, or c1;

- b2-1 belongs to the group with medium frequency of depolarization of a1, b1, or c1;
- c2-1 belongs to the group with low frequency of depolarization of a1, b1, or c1.

Each subgroup of the 59,049 subgroups is numbered and grouped according to their characteristics.

1–59049	I, II, or III	A1, B$_1$, or C1	A2, B2, or C2	D, E, or F	G, H, or I	J1, K1, or L1	J2, K2, or L2	a, b, or c	a1, b$_1$, or c1	a2, b2, or c2

A prokaryotic species will have these marking according to a fictitious example in the below table.

2,345	I	A1	C2	F	I	J1	J2	a	B$_1$	c2

The 59,049 groups can store millions of prokaryotic cells. The prokaryotic cells can be further grouped according to

1. shape (17 groups: see the different shapes of prokaryotic cells);
2. gram stain (3 groups: gram-positive, gram-negative, and prokaryotic cells that are not stained by the gram; 17 x 3);
3. different enzymes.

The number, the characteristics, and name of species can be stored in computers that can find species according to their number and characteristics.

Nobody knows how many prokaryotic species exist and why prokaryotic cells live together in different ecosystems. This practical taxonomy will enable research of prokaryotic ecosystems and identify principles that enable an existence of prokaryotic cells in different ecosystems. The unknown principle will enable the designing of different prokaryotic ecosystems for microbiological needs.

Methods

Methods are built according to the practical taxonomy of prokaryotic cells. See appendix XLI, which explains the methods of practical classification. The methods are inseparable of the new taxonomy because they provide necessary information for it.

CHAPTER 16
STRUCTURES OF EUKARYOTIC CELLS ACCORDING TO THE NEW APPROACH

Attracted complex cylinders build the nucleus, cytoplasm with organelles, and cell membranes of eukaryotic cells (appendix XVIII). Complex cylinders consist of four protein cylinders inside each other. Four protein cylinders are united in two protein cylinders in the middle of a complex cylinders. The first two cylinders from the outside of a complex cylinder are united in the outside cylinder. The second two cylinders from the inside of a complex cylinder are united in the inside cylinder. The outside and the inside cylinder build an arm of the chromosome. The fours cylinders in the opposite sides of the complex cylinders build the different cells' organelles.

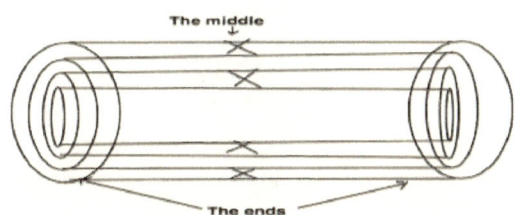

The diagram presents four cylinders. The two outside cylinders are united in the outside cylinder. The two inside cylinders are united in the inside cylinder.

The outermost cylinder is the first; the innermost is the fourth. The cylinders consist of four strings, two different synthetic and two different decomposition strings. The first and second cylinders consist of synthetic and decomposition strings. The third and fourth cylinders consist of the other synthetic and decomposition strings.

The synthetic strings build the first and third cylinders on one side of the complex cylinder, and the second and fourth cylinder do so on the opposite side. The decomposition strings build also the first and third cylinders on one side of the complex cylinder and the second and fourth cylinder do so on the opposite side. The synthetic and decomposition strings unify in the middle of the complex cylinder. The outside cylinder in the middle of the complex cylinder consists of the synthetic and decomposition strings, whereas the outside cylinder in the middle of the complex cylinder consists of the other synthetic and decomposition strings.

The different strings consist of decomposition and synthetic protein spirals, which consist of polypeptides. Metabolism occurs in the strings of eukaryotic cells, the same as in the strings of prokaryotic cells. The specific organelles of eukaryotic cells have specific and separate metabolic functions according to the literature, so they are explained as separate entities. According to the new approach, the specific organelles have similar functions that enable metabolism, growth, and the division of eukaryotic cells. The specific organelles are explained in more detail in chapter 17.

Complex cylinders are rolled around nucleolus in the nucleus. They stretch from a pole of a cell to another one building cytoplasm and its organelles.

Complex cylinders roll up around nucleolus into the nucleus and they get out and get in it via the Golgi apparatus. They stretch from a pole of a eukaryotic cell to another one, building cytoplasm with its different organelles. The complex cylinders are organized in pairs that interchange parts of protein and nucleic acid spirals at opposite poles of eukaryotic cells, the same as the pairs of filaments in prokaryotic cells. Each pair of complex cylinders consists of one mother and one father complex cylinder. The diagram shows one pair of the complex cylinders that build the different organelles.

Cell membrane, endoplasmic reticulum, microtubules, mitochondria and mitotic or meiotic filament	Short and the long arm of the maternal chromosome	Cell membrane, endoplasmic reticulum, microtubules, mitochondria and mitotic or meiotic filament
Cell membrane, endoplasmic reticulum, microtubules, mitochondria and mitotic or meiotic filament	Short and the long arm of the paternal chromosome	Cell membrane, endoplasmic reticulum, microtubules, mitochondria and mitotic or meiotic filament

Complex cylinders consist of long tubes that look like sausages of endoplasmic reticulum and mitochondria of different sizes and lengths. Complex cylinders appear as sausages due to torsion caused by the strings' spiral movement. The long tubes of the complex cylinders attract each other into compact cytoplasm with different organelles.

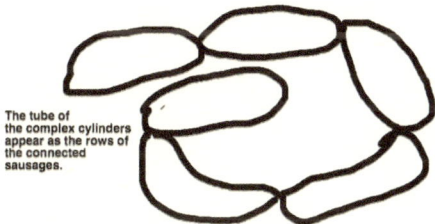

The tube of the complex cylinders appear as the rows of the connected sausages.

Complex cylinders consist of the structures similar to connected sausages. Electronic microscopes cannot reveal the complex cylinders due to their attractions among themselves.

Coenzymes in Polypeptides of Protein Spirals of Strings in Complex Cylinders

Prokaryotic cells of different species have peptidoglycan molecules with the same positions of coenzymes in polypeptides that repeat along the protein spirals of the strings to produce them.

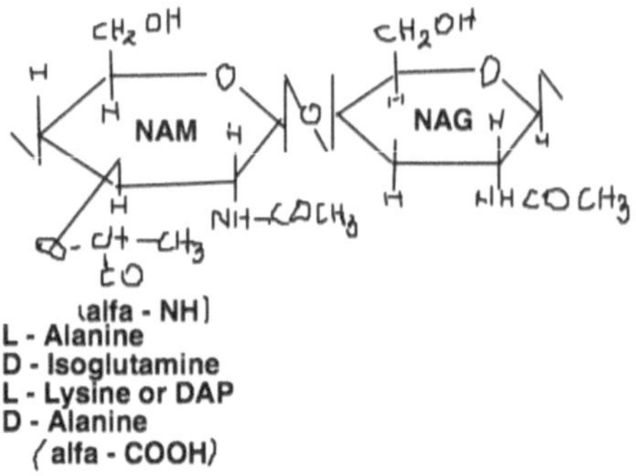

NAM acids (NAM) have the short polypeptides. NAM and NAG molecules repeat alternatively in their chains.

Eukaryotic cells have hyaluronic acids and the same positions of coenzymes in polypeptides that repeat along the protein spirals of the strings to produce them.

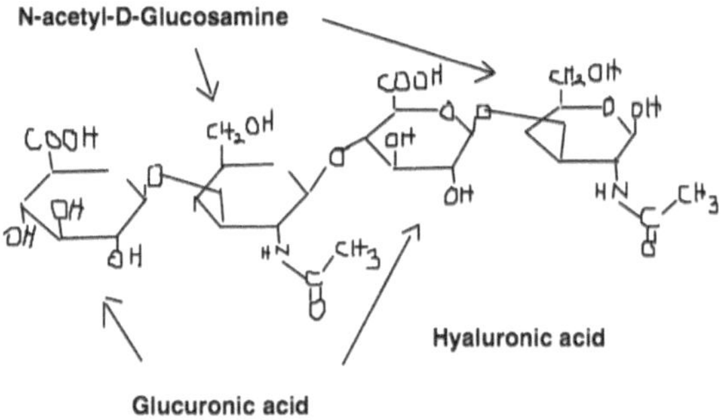

Glucosamine acid and NAG repeat alternatively in the chain of hyaluronic acid.

Filaments of prokaryotic cells and complex cylinders of eukaryotic cells produce different quantities of the same simple metabolites simultaneously. They have polypeptides with different quantities of coenzymes that attract metabolites. The positions of the specific coenzymes in the specific polypeptides of the different filaments of the different prokaryotic cells are the same as those of eukaryotic cells. The molecules of peptidoglycan are similar to those of hyaluronic acid produced by eukaryotic cells; polypeptides of eukaryotic and prokaryotic cells have different concentrations of the same coenzymes that produce glucuronic acid of hyaluronic acid and NAM of peptidoglycan molecules.

Polypeptide 1	Polypeptide 2	Polypeptide 3	Polypeptide 4	Polypeptide 5	Polypeptide 6	Polypeptide 7	Polypeptide 8
ATP a. (--)	B_1, Biotin, CoA c. (+)	B_6, Fol. e. (-)	NAD, FAD g. (+)	ATP i. (--)	B_1, biotin, CoA k. (+)	B_6, Fol. K m. (-)	NAD, FAD o. (+)
NAD, FAD b. (+)	K d. (-)	B_{12} f. (+)	ATP h. (--)	NAD, FAD j. (+)	K, Biotin l. (-)	B_{12} n. (+)	ATP p. (--)

Coenzymes as bipolar molecules have electrical charges that attract opposite electric charges of bipolar metabolites	Metabolites as bipolar molecules have electrical charges that attract opposite electric charges of bipolar coenzymes
ATP+	PO4--
FAD-, NAD-	H+
B_1+	pyruvate+
biotin+	different acids of the Krebs cycle+
CoA	acetyl+
K vitamins	fatty acids-
B_6	amino acids-
folic acid	acetyl-
B_{12}	Fe++

The eight polypeptides with specific coenzymes attract four groups of carbon atoms of the glucose ring at the polysaccharide molecules. The specific polypeptides in the positions 1 and 5, 2 and 6, 3 and 7, and 4 and 8 have the same structures of the specific coenzymes but with different quantities. Polypeptides with specific coenzymes and their different quantities enable them to produce N-acetyl-D-glucosamine and D-glucuronic acid at eukaryotic cells or N-acetyl glucosamine and NAM at prokaryotic cells.

Other differences exist between the polypeptides of eukaryotic and prokaryotic cells. Polypeptides of eukaryotic cells are bigger and have attraction among them greater than that of prokaryotic cells. Polypeptides of eukaryotic cells have nonprotein hormones that enable stronger attraction among polypeptides whereas the polypeptides of prokaryotic cells do not.

The longer specific polypeptides and the stronger attraction among them enable longer and thicker strings of eukaryotic cells than the strings of eukaryotic cells have. Longer strings are necessary for building bigger structures of complex cylinders than filaments.

The specific nonprotein hormones with specific ions as aldosterone—(Na+), adrenalin—(Cl –), progesterone—(Cu+), thyroxin—(I-), estrogen—(Zn++) and others help better attractions among polypeptides and longer protein spirals of the longer and the thicker strings of complex cylinders.

The eight polypeptides at prokaryotic cells do not have nonprotein hormones and therefore have shorter strings of filaments and shorter filaments than do the complex cylinders of eukaryotic cells. This is an essential difference between the polypeptides of eukaryotic cells and those of prokaryotic cells. Nonprotein hormones with ions are arranged at polypeptides so they can attract each other as opposite electrical charges. The possible connections of the nonprotein hormones, ions, and polypeptides are not still theoretically handled in this text. The logic of this approach is that stronger attraction of polypeptides at polypeptide and protein spirals enable the creation of longer protein spirals in eukaryotic cells that enable the longer strings necessary for building complex cylinders.

Eukaryotic cells of species that do not produce hyaluronic acids have most likely the same positions of coenzymes at polypeptides because they have the same simple and complex metabolites as do eukaryotic cells that produce hyaluronic acids.

Genes of eukaryotic cells are organized in groups in nucleic acid spirals the same as with prokaryotic cells. They repeat along the nucleic acid spirals to synthesize polypeptides of protein spirals the same as with prokaryotic cells.

In summary, the eight specific polypeptides with specific coenzymes and metabolites must repeat to produce N-acetyl-D-glucosamine and D-glucuronic acid of hyaluronic acid in eukaryotic cells and NAG and NAM acid of peptidoglycan molecules at prokaryotic cells.

Differences between Prokaryotic and Eukaryotic Cells

Prokaryotic cells are built from pairs of similar filaments that build the filaments that build their different structures (pili, flagella, cell membrane, plasmids, and nucleoids). Filaments of prokaryotic cells have two strings in a specific organization.

The model of filament has the two strings that have the specific structure of coils and straight parts. Coil parts grow around straight parts whereas straight parts do not grow. The new model of filament is called the Rosalind Franklin model because protein spirals of strings have the organization of ds DNA.[41]

Eukaryotic cells are built from pairs of complex cylinders that build their different structures (nucleolus, nucleus, chromosomes, the Golgi apparatus, endoplasmic reticulum, mitochondria, centrosome body, cell membranes, and others). The complex cylinders are built of four strings with specific organization in the biggest one. The first and second cylinder build the outside cylinder, whereas the third and fourth build the inside cylinder. The first and second cylinders and the third and fourth cylinders have a similar organization of prokaryotic filaments. This sentence will be clear after a short explanation of the hypothetic evolution of eukaryotic cells explained in the below paragraph. Evolution of eukaryotic cells is explained in more details in chapter 19.

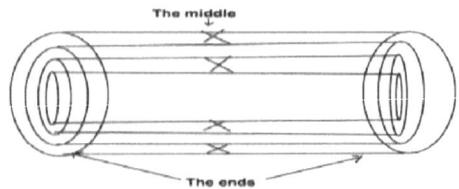

The four strings of the four cylinders from both sides are unified into the two cylinders in the middle of the complex cylinder.

Complex cylinders of eukaryotic cells are much bigger and longer than filaments of prokaryotic cells. Prokaryotic and eukaryotic cells were born from archaic prokaryotic and eukaryotic cells that had archaic strings of different qualities and lengths. Archaic strings of archaic complex cylinders of archaic eukaryotic cells were longer than archaic strings of archaic filaments of archaic prokaryotic cells. Long protein spirals of long archaic strings of archaic eukaryotic cells had long polypeptides and nonprotein hormones in a comparison with short protein spirals of short archaic strings of archaic prokaryotic cells that had short polypeptides without nonprotein hormones.

The appearance of nonprotein hormones and longer polypeptides enabled stronger attraction among archaic polypeptides and the creation of longer strings, which enabled the evolution of longer and thicker filaments in complex cylinders with two protein cylinders at their ends and one in the middle.

The predecessors of archaic eukaryotic cells had been archaic complex cylinders with two cylinders at their ends and one in the middle; these had swallowed complex cylinders also with two cylinders at their ends and one in the middle of smaller diameters. Together, they had formed archaic complex cylinders of four cylinders at the ends and two in the middle. Archaic eukaryotic cells had been born by unification of archaic complex cylinders with four cylinders at the ends and two in the middle. Environmental conditions negated archaic forms of prokaryotic and eukaryotic cells during their long evolution.

Strings of prokaryotic and eukaryotic cells are built on decomposition and synthetic protein spirals that act on metabolites simultaneously;

they create unity and conflict of opposites. These spirals are made up of polypeptides and amino acids that attract coenzymes that attract metabolites as ions.

Prokaryotic and eukaryotic cells have similar organization of coenzymes in protein spirals' polypeptides established by the similar structures of peptidoglycan, hyaluronic acid, and other molecules produced by metabolism. The complex molecules look for the same organization as specific coenzymes among attracted polypeptides for their synthesis and decomposition. Coenzymes in polypeptides have different quantities that enable the production of different quantities of metabolites and complex molecules and with them different qualities of prokaryotic and eukaryotic cells. This is according to the law of the passage of quantitative to qualitative changes based on universal laws of nature.

Polypeptides of eukaryotic cells have nonprotein hormones that are most likely at the ends of the polypeptides. Nonprotein hormones (estrogen, progesterone, adrenalin, and other) attract ions (Zn++, Cu+, I-, Cl, and others). Nonprotein hormones enable stronger attraction among polypeptides of the long strings of eukaryotic cells by opposite attractions among their ions. Nonprotein hormones are not present in prokaryotic cells.

Strings of eukaryotic cells are thicker and longer than those of prokaryotic cells due to explained reasons.

Prokaryotic and eukaryotic cells' strings have ds DNA with negative and positive strands. The positive strands bring different genes responsible for biosynthesis of polypeptides. The genes are organized in groups with two subgroups of similar genes; the subgroups permeate each other along the positive strands of ds DNA.

1	1'	2	2'	3	3'	4	4'	5	5'	6	6'	7	7'	8	8'

Eight specific genes of decomposition protein spirals permeate the eight specific genes of synthetic protein spirals along positive strands of ds DNA.

One subgroup of genes in the different groups produces a group of polypeptides of synthetic protein spirals that synthesize and decompose metabolites. Another subgroup produces polypeptides of decomposition protein spirals that decompose and synthetize different metabolites (synthetic protein spiral synthetize more carbohydrates, while decomposition protein spiral decompose more carbohydrate – those protein spirals got names according to metabolism of carbohydrates). The groups of genes repeat along ds DNA of the strings of filaments of prokaryotic cells and the complex cylinders of eukaryotic ones.

The nucleic acids of a prokaryotic cell's filaments have different quantities of genes due to their lengths. Prokaryotic cells with more groups of genes can produce more phenotypes.

Groups of genes of eukaryotic cells are organized in bigger groups that do not have the same groups. They repeat also along the positive strands of ds DNA of the strings of complex cylinders. The nucleic acids of strings of cylinders of eukaryotic cells have different quantities of the same bigger groups because the different complex cylinders have different lengths.

Different groups of genes produce synthetic and decomposition protein spirals in strings of prokaryotic and eukaryotic cells. The quantity of genes determines the length of the strings because bigger quantities of genes make bigger quantities of polypeptides and their protein spirals. The complex organism has more different eukaryotic cells with the same nucleic acids; different genes organized in bigger groups produce different eukaryotic cells, which are produced during stem cell differentiation.

Eukaryotic cells' nucleic acid spirals are much longer than those of prokaryotic cells and have more groups of genes than do the nucleic acids of prokaryotic cells.

Prokaryotic and eukaryotic cells synthesize polypeptides and genes and assemble protein spirals and nucleic acids in the same way. Eukaryotic cells have bigger ribosomes than do prokaryotic cells because polypeptides of eukaryotic cells are longer than those of prokaryotic cells. Longer polypeptides make bigger protein spirals and their ribosomes. Pairs of filaments of prokaryotic cells and pairs of complex cylinders of eukaryotic cells interchange parts of protein and nucleic acid spirals, which enables growth of strings and metabolism. Pairs of filaments of prokaryotic cells and pairs of complex cylinders of eukaryotic cells create unity and conflict of opposites. Prokaryotic and eukaryotic cells metabolize carbohydrates, fats, amino acids, coenzymes, nucleotides, and other molecules. Eukaryotic cells produce nonprotein hormones, cholesterol, and molecules similar to cholesterol, whereas prokaryotic cells do not.

Prokaryotic and eukaryotic cells polarize and depolarize alternatively producing movements of protein spirals and metabolites as ions, which produce their bioelectricity. Eukaryotic cells have more bioelectricity than do prokaryotic cells.

Prokaryotic cells divide by binary fission whereas eukaryotic cells divide by mitosis. Filament division of prokaryotic cells causes binary fission whereas division of complex cylinders in eukaryotic cells causes mitosis and meiosis. Filaments and cylinders divide by the chewing gum phenomenon, which is possible only if filaments and cylinders of eukaryotic cells have specific organization of strings in their structures.

Eukaryotic cells of complex organisms undergo meiosis, which enables mixing of nucleic acids and their genes during the crossing over and reduction of the diploid number of the complex cylinders and their chromosomes into the haploid number in the predecessors of sex cells. The haploid number of cylinders and their chromosomes in sex cells enables associations of their nucleic acids with their genes during their unification into zygote cells. Different combinations of similar genes produce new organisms with different phenotypes (qualities) subject to natural selection and environmental conditions.

Different cells transform into malignant (cancer) ones in complex multicellular organisms. Cells become malignant by gene mutations that produce new qualities of polypeptides, which causes poor attraction among polypeptides, broken or abnormal complex cylinders, cytoplasm, nucleus, metabolism, and mitosis. Such malignant transformations are explained in chapter 22.

Dialectic laws govern characteristics of prokaryotic and eukaryotic cells including malignant ones. Prokaryotic and eukaryotic cells live in colonies that enable them to better survive and evolve in different environments.

In summary, the differences between prokaryotic and eukaryotic cells in essence are not big despite their having different structures, shapes, and sizes.

Chapter 17
Complex Cylinders and Organelles

Complex cylinders build cytoplasm and nuclei with different organelles. The Golgi apparatus is a place of branching of complex cylinders that get out from a nucleus and get in it building the cytoplasm's and the nucleus' organelles (appendix XIX). The Golgi apparatus disappears during mitosis and meiosis the same way a nucleus does. Some organelles appear and disappear during a cell's life cycle because of coiling and uncoiling of complex cylinders' strings, explained in chapter 20.

The cylinders create different structures of a cell. The first cylinders build the rough and the smooth endoplasmic reticulum (appendix XX). The strings of the first cylinder attract each other, make compact membranes, and build cytoplasmic membranes if the complex cylinders are on the periphery of eukaryotic cells (appendix XXI). The growing parts of the decomposition or the synthetic strings build the spirals of the first cylinder, the biggest.

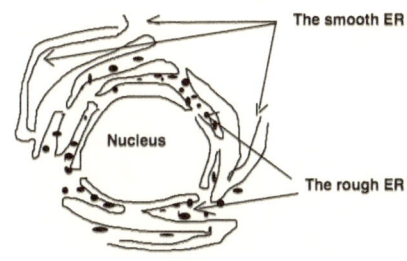

Endoplasmic reticulum

Electronic images do not show the complex cylinders inside cytoplasm because the complex cylinders attract each other in the compact cytoplasm that has a net of the tubules with the rough and the smooth ER.

The second cylinders build the microtubules and filaments (appendix XXII). The strings of the second cylinders attract each other, and they do not build membranes but only spirals in the spiral (tubules) that connects the first and third cylinders.

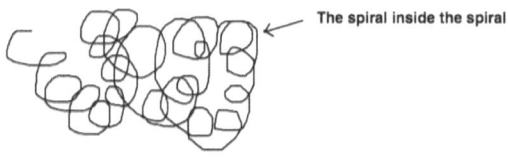

The strings of the second cylinders do not build the compact membranes but only the spirals inside the spirals that connect the first and the third cylinders.

Nongrowing parts of decomposition or synthetic strings build spirals of the second cylinder. The second cylinder with microtubules connects endoplasmic reticulum and mitochondria.

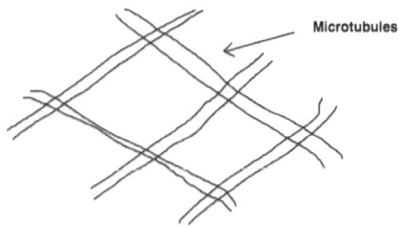

Electron images do not show the spiral in the spiral but only microtubules as the net of the tubules without any structures.

The nongrowing parts of the strings that are more coagulated make the mash of filaments in electron microscope images; the images cannot observe the cells' structures because their proteins are coagulated.

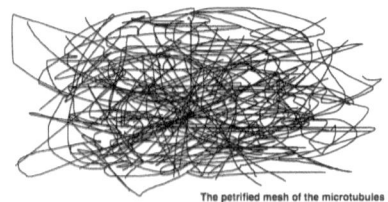

The petrified mesh of the microtubules

Electron images show a petrified mash of strings (filaments) due to their coagulated proteins. The strings have protein spirals in protein spirals as strings of prokaryotic cells.

The third cylinders build the membranes of mitochondria (appendix XXIII). The strings of the third cylinders also build compact membranes. The growing parts of decomposition or synthetic strings build the spirals of the third cylinder.

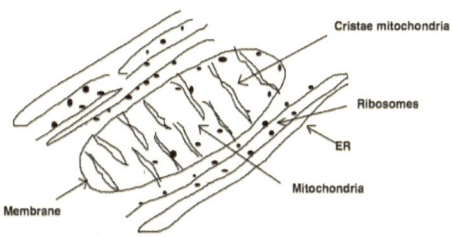

Electron images show a mitochondrion with a membrane and cristae mitochondria in it. The membrane is built from growing parts of the third cylinders. Cristae mitochondria are built from nongrowing parts of strings that build the fourth cylinders. The cylindrical shape of a mitochondrion cannot be seen in the electronic images because the cell with its organelles is cut into a plane before its examination by the electron microscope.

The fourth cylinders build cristae mitochondria. The strings of the fourth cylinders also build compact membranes that are wrinkled in complex cylinders due to lack of space. The nongrowing parts of decomposition or synthetic strings build the fourth cylinder's spirals.

Chromosomes

By unifying synthetic and decomposition strings, the first and second cylinders build the outside cylinder of the chromosome in the middle of the complex cylinder. The third and fourth cylinders make the inside cylinder in the same way. The outside and inside cylinders of complex cylinders attract each other and make chromosomes that get shape during mitosis and meiosis. The chromosomes are more complex than they are shown in the literature (appendix XXIV). Chromosomes can be seen before and during cell division; their strings of chromosomes have the same metabolism as strings of other organelles.

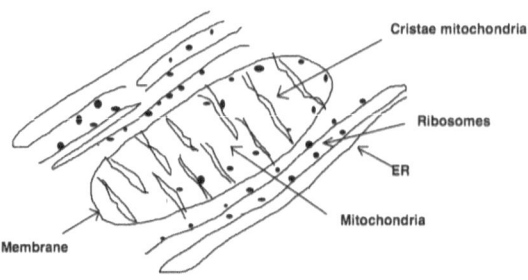

Chromosomes connect the cytoplasm's organelles by unknown mechanisms according to contemporary literature. Eukaryotic cells of plant and animal species have different numbers of chromosomes organized in pairs (2n). The arms of chromosomes have similar positions of similar genes. An arm of the chromosome has its mother's origin while another has its father's origin. This fact is also accepted by the new approach.

Different organisms	Number of chromosomes
rye	10
maize	14
tobacco	48
bread wheat	42
fruit fly	8
human	46
earthworm	36
gorilla	48

Chromosomes are connected by pairs of complex cylinders that interchange parts of protein spirals and nucleic acids in cytoplasm. A chromosome consists of short and long arms connected by centromeres.

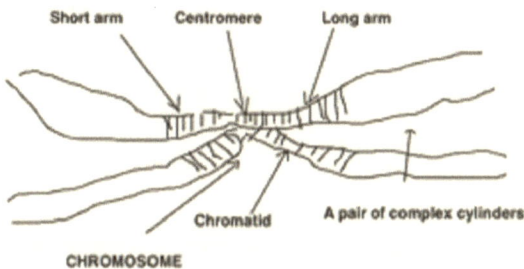

CHROMOSOME

Diagram shows a condensed and metaphase eukaryotic chromosome connected with a pair of complex cylinders. The chromosome consists of chromatid (one of the two identical parts of the chromosome after S phase), centromere (the point where the two chromatids touch each other), short arm, and long arm.

The densely coiled strings of complex cylinders make the short and long arms of chromosomes. One short and one long arm make an arm of a chromosome of the paternal or the maternal side during metaphase at a centromere, a place of torsions of complex cylinders during mitosis. This torsion phenomenon is explained in mitosis according to the new approach.

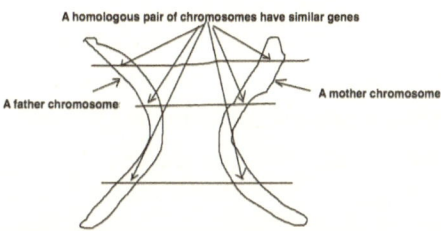

A homologous pair of chromosomes. The duplications of chromosomes during mitosis are explained by duplication of chromatids according to the literature. The literature cannot explain duplication of chromosomes during mitosis and meiosis in a logical way because chromosomes exist independent of other structures of cytoplasm. The duplication of chromosomes is explained in the new approach of mitosis in chapter 20.

Homologous chromosomes and their complex cylinders are not attracted firmly along their lengths during mitosis but only during meiosis; such firm attraction along their lengths enables crossing over and reduction of chromosomes during meiosis. Both phenomena are explained in mitosis and meiosis according to the new approach.

Nucleolus and Nucleus

A nucleolus is a place of bending of the complex cylinders and their rolling immediately after mitosis and meiosis according to the new approach.

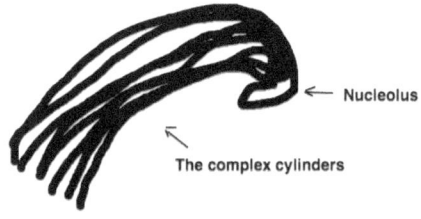

← Nucleolus

The complex cylinders

A nucleolus a place of bending of the complex cylinders immediately after mitosis and meiosis.

Bending of new complex cylinders before rolling

A group of bent complex cylinders is rolled into nucleolus, nucleus (appendix XXV), the Golgi apparatus, and organelles of cytoplasm. The rolling starts at the bent site of the nucleolus and finishes with the centrosome body immediately after mitosis. The middle parts of the complex cylinders with two cylinders make the nucleus; the rest of the complex cylinders with four cylinders make cytoplasm with organelles.

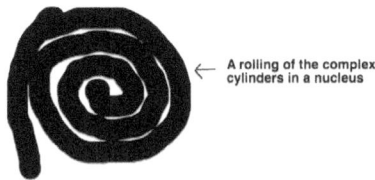

← A rolling of the complex cylinders in a nucleus

A snail resembles the rolling of a complex cylinder into nucleolus, nucleus, the Golgi apparatus, and other organelles of cytoplasm. The rolling cannot be seen under electron microscopes because it occurs in a short time immediately after mitosis.

Strings of Complex Cylinders and Organelles

Complex cylinders consist of four different strings that consist of decomposition and synthetic protein spirals. The protein spirals consist of polypeptides.

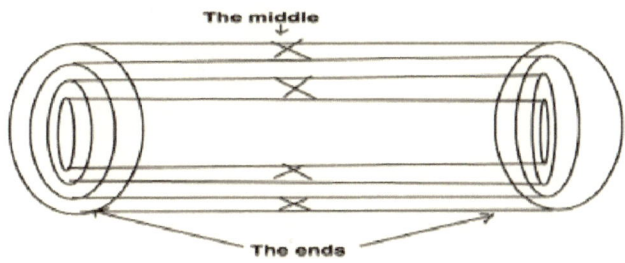

Four strings of the four cylinders from both sides are unified into the two cylinders in the middle of the complex cylinder.

Complex cylinders exist as pairs; the specific positions of their ends enable the interchange of parts of protein and nucleic acid spirals. The diagram below shows the different strings of the complex cylinder with the movements of the different protein spirals, the different nucleic acids, and the new protein spirals.

Synthetic string (endoplasmic reticulum)

Synthetic protein spiral, ds DNA and tRNA >>> <-ss DNA of dsDNA	Synthetic protein spiral and ds DNA >>> <<< ds DNA
<<< New synthetic protein spirals, ds DNA, and tRNA on decomposition protein spirals with +ss DNA of dsDNA <<<	Decomposition protein spirals with +ss DNA of dsDNA <<<

| Growing part | Nongrowing part |

Decomposition string (microtubules)

<<< Decomposition protein spiral and ds DNA ds DNA >>>	<<< Decomposition protein spiral, ds DNA, and tRNA >-ss DNA of dsDNA
Synthetic protein spirals with ds DNA >>>	New decomposition protein spirals, ds DNA and tRNA on synthetic protein spirals +ss DNA of ds DNA >>>

Nongrowing part   Growing part

The third and the fourth cylinders of the complex cylinder have these positions of synthetic and decomposition strings.

Synthetic string (membrane of mitochondria)

Synthetic protein spiral, ds DNA, and tRNA >>> <-ss DNA of dsDNA	Synthetic protein spiral and ds DNA >>> <<< ds DNA
<<< New synthetic protein spirals, ds DNA, and tRNA on the decomposition protein spirals with +ss DNA of dsDNA <<<	Decomposition protein spirals with +ss DNA of dsDNA <<<

Growing part   Nongrowing part

Decomposition string (cristae of mitochondria)

<<< Decomposition protein spiral and ds DNA ds DNA >>>	<<< Decomposition protein spiral, ds DNA, and tRNA >-ss DNA of dsDNA
Synthetic protein spirals with ds DNA >>>	New decomposition protein spirals, ds DNA, and tRNA on synthetic protein spirals +ss DNA of ds DNA >>>

Nongrowing part   Growing part

Synthetic and decomposition strings that build the first and second cylinders are not the same as the synthetic and decomposition strings that build the third and fourth cylinders. This is repeated due to its importance.

Pairs of cylinders interchange parts of protein spirals and nucleic acids as pairs of filaments of prokaryotic cells. The cylinder pairs are also called homologous complex cylinders. Homologous chromosomes are parts of the homologous complex cylinders; they can be seen during mitosis and meiosis.

Eukaryotic cells consist of pairs of homologous complex cylinders and chromosomes that are not the same. The decomposition and synthetic strings of cylinders must have the same synthetic and decomposition protein spirals to interchange parts of protein spirals and nucleic acids. The cylinders must have the growing and nongrowing parts of decomposition and synthetic strings on the same side to exchange the parts.

Polypeptide synthesis occurs in the growing strings of the cylinders. Ribosomes are temporary structures that appear during polypeptide biosynthesis according to the new approach. Ribosomes appear in groups as polyribosomes in endoplasmic reticulum or cell membrane, microtubules, the membrane of mitochondria and crista mitochondria and less in the Golgi apparatus and nucleus. They are seen under electron microscopes most commonly below endoplasmic reticulum, the mitochondrial membrane, and in the nucleus. There is no logical explanation for their existence in a cell structure if the strings do not exist with protein spirals. Polyribosomes cannot be separated as sole entities because there is no explanation for how specific protein produced in polyribosomes reaches specific places and builds specific structures. Ribosomes exist as separate entities in electron images because the electron images are images of the coagulated and disrupted structures.

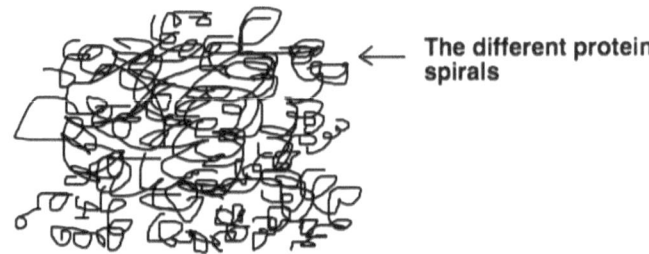

← The different protein spirals

The contemporary model of a ribosome looks like a bunch of the different protein spirals

The contemporary model of a ribosome does not explain its origin and function.

The model of a eukaryotic cell according to the contemporary literature does not reveal any connection among the organelles and their existence as a unified cell system.

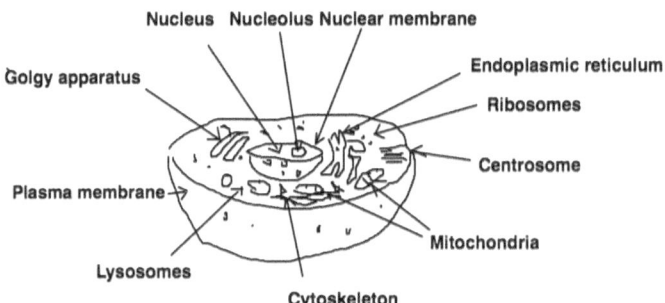

The model of a cell produced by the electron images of the cells is irrational because the organelles swim inside cytoplasm without any connections.

The electron images of eukaryotic cells do not reveal any connections among the organelles because the cells were killed before their making. The protein structures of the cells are coagulated and disrupted. The electron images of prokaryotic and eukaryotic cells produced the wrong ideas about their structures and functions. The electron images of eukaryotic cells do not reflect the real nature of the cell's organelles.

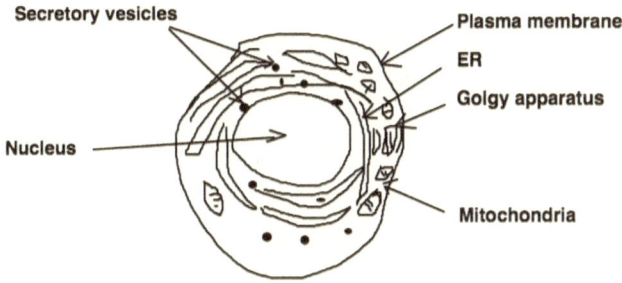

Electron images of a eukaryotic cell show the organelles without out any connections.

When squeezed, complex cylinders make a double membrane of the nucleus. The places of narrowing and touching of the complex cylinders make an impression of pores in the electron images of the nucleus.

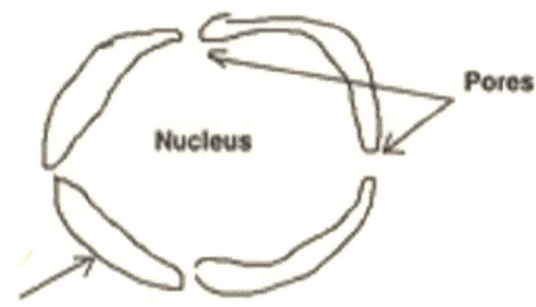

Electron image of a nucleus showing the nuclear pores and the double membrane.

The pores are not places where protein and nucleic acids get out according to the new approach. The movements of the protein and nucleic acids in the strings of eukaryotic cells take place in the same way as the movements of protein and nucleic acids in the strings of prokaryotic cells. Nuclei consist of the rolled complex cylinders that attract each other.

Rolled complex cylinders are also not seen in electron images of the nucleus. The complex cylinders in nuclei make the chromatin net during interphase and metaphase. The complex cylinders are the longest immediately after mitosis and the shortest during anaphase. The transformation of the complex cylinders into chromosomes is explained in more detail in chapter 20.

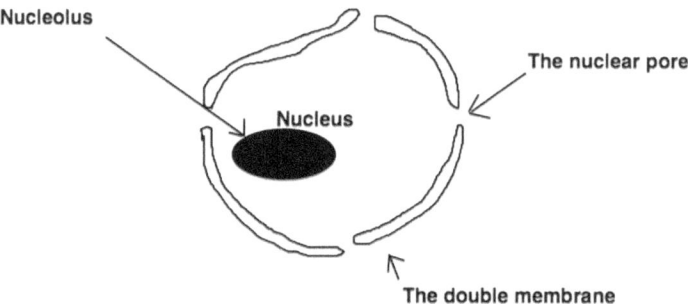

Electron images show nucleolus inside of nucleus like a dark stain.

The squeezed complex cylinders of the four cylinders of the Golgi apparatus get into or out of the nucleus as two unified branches. The two unified branches of the Golgi apparatus roll into different organelles of cytoplasm.

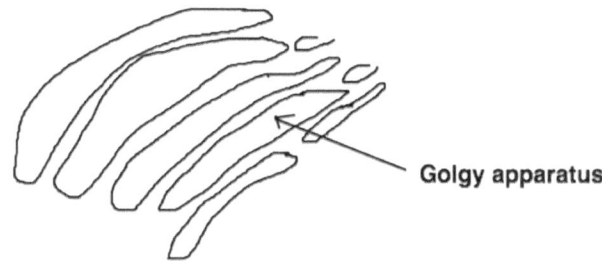

Electron images of the Golgi apparatus show a group of long cisterns.

The ends of the strings of the two groups of the complex cylinders have different lengths due to the rolling of complex cylinders like a snowball.

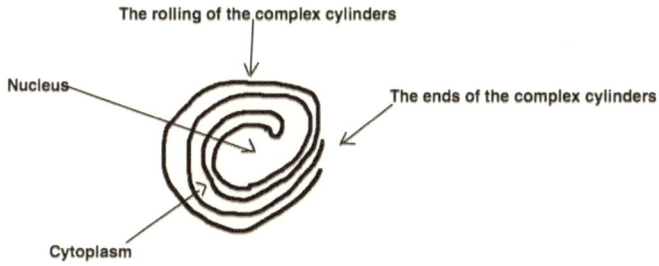

Consequences of rolling complex cylinders are ends of different lengths.

The ends of the strings of the two groups of the complex cylinders with different lengths swirl and unify into two centrioles of a centrosome body. Two centrioles make a position of the right angle due to the different lengths of the two groups of the complex cylinders.

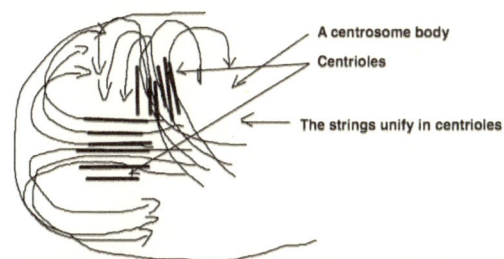

The proof for this approach is that the electron images of a centrosome body show the position of two centrioles in the right angle.

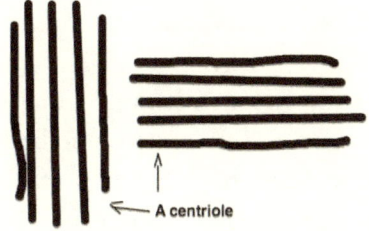

Electron images show two centrioles at right angles in a centrosome body.

The proof for this approach is also that electronic images show the immersing of the strings in a centriole.

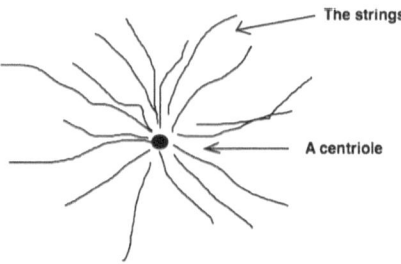

Electron images of a centriole show the strings that branch from a centriole.

The swirling of strings of the complex cylinders of a group of the complex cylinders can be seen in the structures of a centriole (appendix XXVI).

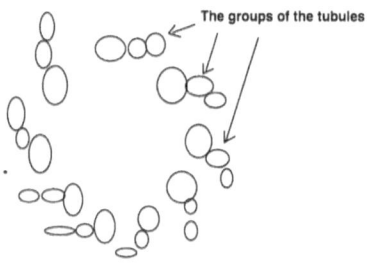

The electron images of a centriole show the nine groups of the strings that have swirling positions.

A centriole consists of nine groups of three strings with swirling positions and two in its center according to electronic images. The ends of the strings of complex cylinders form the centrioles due to their attractions.

A centrosome body is formed immediately after mitosis due to attraction between centrioles. A centrosome body separates into two centrioles during late interphase before mitosis or meiosis due to reducing the lengths of complex cylinders and their transformation into chromosomes and mitotic strings. The complex cylinders transform into mitotic strings with chromosomes in mitotic spindles between two centrioles. The below pictures show separations of the centrosome body into the two centrioles.[8] The pictures do not contradict the new approach explained here.

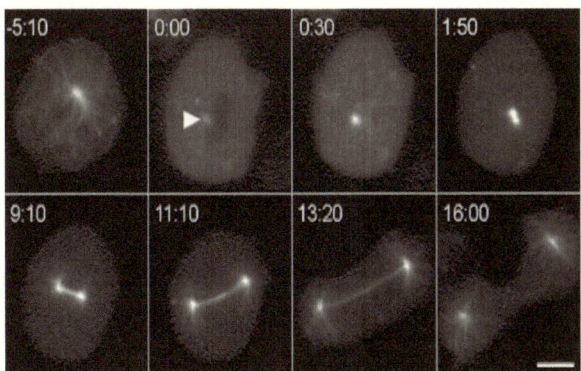

Movements of the two groups of the complex cylinders into two opposite poles of a cell are due to reducing of lengths of the complex cylinders that transform into strings with chromosomes in the mitotic spindle between centrioles. The sequence shows cell division in dictyostelium. The author of the pictures is Michael Koonce PhD, Chief, Cellular and Molecular Basis of Disease and Director of Division of Translational Medicine, Wadsworth Centre.

The centrioles look like stars in the mitotic spindle in the electronic images of cells in mitosis. The centrioles as stars are explained by the unification of strings of complex cylinders transformed into mitotic spindles and chromosomes.

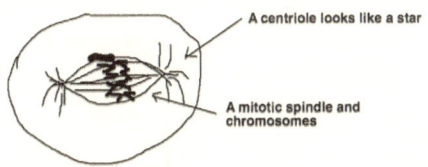

Electron images of a cell in mitosis show centrioles as the two stars connected with strings of the mitotic spindle.

Strings that get out of the centrioles bend and finish as free strings in cytoplasm; they release and exchange parts of protein and nucleic acids spirals in lysosomes. These parts accumulate outside complex cylinders forming sacks among endoplasmic reticulum; the sacks are known as lysosomes in the literature and here as well.

The endoplasmic reticulum (ER) is a system of channels with walls built by the first cylinders of the complex cylinders according to the new approach. According to the literature, "The endoplasmic reticulum serves many general functions, including the folding of protein molecules in sacs called cisternae and the transport of synthesized proteins in vesicles to the Golgi apparatus."[9] "The Golgi apparatus is a major collection and dispatch station of protein products received from the endoplasmic reticulum (ER)."[10] The Golgi apparatus is a supercomputer that labels items and send them to different parts of the cell without any question as to how according to the literature. Actually, contemporary literature does not have any logical explanation of how proteins are distributed from the Golgi apparatus to the different parts of a cell. This irrational approach had to be replaced with the new approach that at least gives a logical explanation.

The ER does not transfer the newly synthesized proteins from its walls to the Golgi apparatus; rather, it accumulation part of protein and nucleic acid spirals. The Golgi apparatus is a place where complex cylinders of a nucleus branch in two parts of cytoplasm and build different organelles. Proteins as protein spirals move through the strings of the cylinders by spiral movements. They get out and interchange at their endings in lysosomes. How the strings interchange parts of protein and nucleic acid spirals was explained at prokaryotic cell.

ER, which consists of complex cylinders, has two parts—smooth without ribosomes and rough. Smooth ER is close to the nucleus because tRNAs with amino acids are spent during the biosynthesis of polypeptides. Lysosomes belong to the endoplasmic reticulum and are places where strings of pairs of complex cylinders interchange parts of protein and nucleic acid spirals. The parts are accumulated and attract metabolites by their coenzymes. The parts also have the functions of different enzymes; they decompose molecules, viruses, and bacteria.

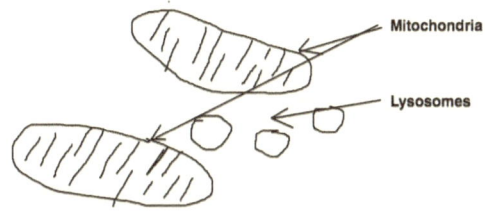

The electron images show lysosomes close to mitochondria due to breaking of complex cylinders. They can degrade the structures of complex cylinders including mitochondria because they contain the parts of the protein spirals that have different enzymatic properties.

Strings of complex cylinders get out from centrosome bodies and build flagella, cilia, dendrites of nerve cells, tails of spermatozoids, and similar structures. Eukaryotic cells move by flagella and interchange parts of protein and nucleic acid spirals with the same cells outside the cells. Flagella of prokaryotic cells consist of filaments whereas flagella of eukaryotic cells consist of groups of the strings.

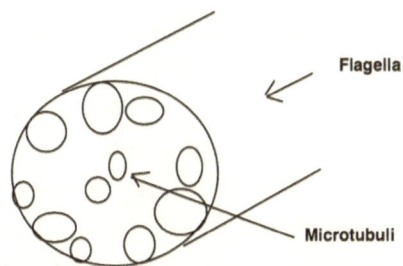

Flagella of eukaryotic cells consist of the associated strings that are here marked as microtubules.

If strings of complex cylinders get out from eukaryotic cells in small and thin associations, they form cilia, which also release parts of protein and nucleic acid spirals. The specific cells with the cilia can interchange these parts. If complex cylinders get out from eukaryotic cells with big associations, they form dendrites and axons of nerve cells.

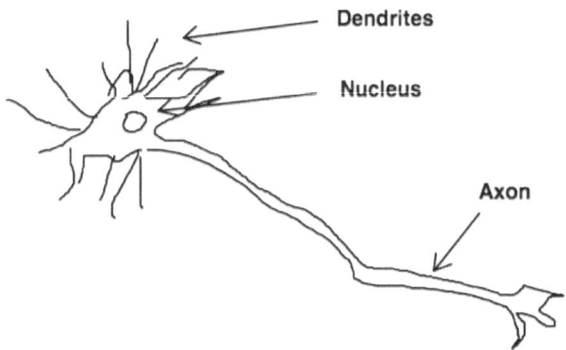

Axons and dendrites of nerve cells are associations of strings according to the new approach.

The strings synthesize protein and nucleic acid spirals in themselves according to the new approach are explained at prokaryotic cells. Ribosomes are produced also in strings of eukaryotic cells. Ribosomes in eukaryotic cells are bigger than those in prokaryotic cells because protein spirals of eukaryotic cells have longer specific polypeptides than prokaryotic ones do (appendix XXVII).

In 1958, Richard B. Roberts said this about the origin of the name *ribosome*.

During the course of the symposium, a semantic difficulty became apparent. To some of the participants, "microsomes" mean the ribonucleoprotein particles of the microsome fraction contaminated by other protein and lipid material; to others, the microsomes consist of protein and lipid contaminated by particles. The phrase "microsomal particles" does not seem adequate, and "ribonucleoprotein particles of the microsome fraction" is much too awkward. During the meeting, the word "ribosome" was suggested, which has a very satisfactory name and a pleasant sound. The present confusion would be eliminated if "ribosome" were adopted to designate ribonucleoprotein particles in sizes ranging from 35 to 100S.[11]

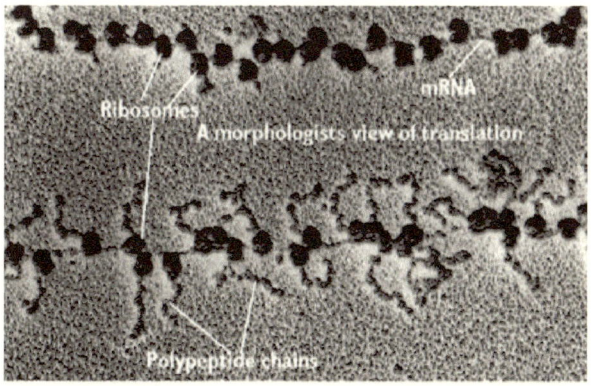

This electron image shows production of polypeptides by ribosomes according to the literature, whose approach is confusing because polypeptides consist of 50–2000 amino acid residues that cannot be big as ribosomes in the picture. The small unit has 33 proteins and 18S RNA (1,900 nucleotides), whereas the large unit has 46 protein and 5S RNA (120 nucleotides), 28S RNA (4,700 nucleotides) and 5.8S RNA (160 nucleotides).[203]

According to the new approach, protein spirals with tRNA that attract amino acids move in one direction while protein spirals with mRNA move in the opposite direction. Protein spirals and their tRNAs with amino acids attracted by mRNA build big units while protein spirals that attract mRNA with amino acids of tRNA build small units. The ribosomes are not permanent structures that last for the life of the cells.

The membrane of eukaryotic cells consists of the fourth cylinders of complex cylinders that are on the periphery. Membranes are not compact because the fourth cylinders consist of synthetic and decomposition strings that attract each other. Strings consist of protein spirals that have polypeptide spirals with fats. Synthetic and decomposition strings with their metabolism function also transport ions of a cell membrane. The contemporary model is dogma because it cannot logically explain this transport or the connections of cells with other structures of cells.

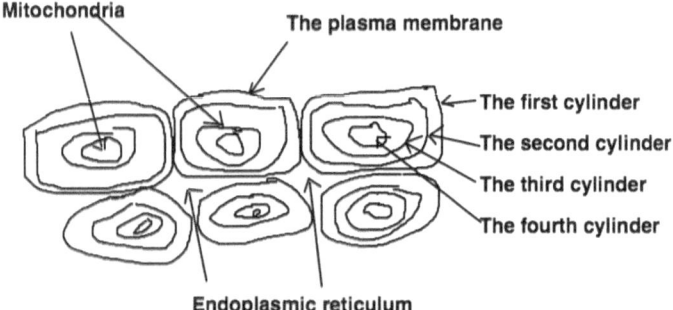

The scheme shows the cut of the complex cylinders. The outside complex cylinders with the fourth cylinders build the plasma membrane according to the new approach. The protein spirals of the fourth cylinders with their metabolites make up the content of the plasma membrane.

CHAPTER 18
MITOSIS, MEIOSIS, AND DIFFERENTIATION OF STEM CELL

Mitosis (division) of eukaryotic cells occurs in two phases—coiling and uncoiling. Coiling occurs during a small part of interphase (G2), prophase, metaphase, and anaphase. The uncoiling phase occurs immediately after mitosis, during G1, S, and a big part of G2 of interphase.

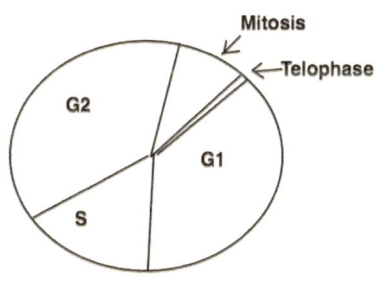

The cell cycle

The peak of the coiling phase occurs during cells' late anaphase and before telophase. The coiled complex cylinders are cut at the peak of the coiling phase due to torsion. The uncoiling phase occurs immediately after a cell divides into daughter cells. The densely coiled strings of the first and third cylinders of the cut complex cylinders in the middle spread as springs into the new second and fourth cylinders, which increase the length of the cut complex cylinders and transform them into incomplete cylinders. These bend because their ends hit the daughter cells and come back along the transformed complex cylinders. The transformed and bent complex cylinders roll into nucleolus, nucleus, and cytoplasm of daughter cells in the onset of G1 of interphase. Everything occurs quickly.

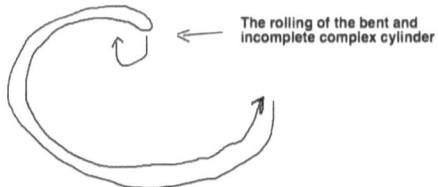

The rolling of the bent and incomplete complex cylinder

The incomplete complex cylinders hit the daughter cell and come back along the transformed complex cylinder

The incomplete complex cylinder rolls into nucleolus, nucleus, and cytoplasm with its organelles.

The strings of the second and fourth cylinders of the cut complex cylinders grow around the new second and fourth cylinders of incomplete complex cylinders as the new first and third cylinders. This occurs in G1 and S interphase. The uncoiling phase is finished when the new second and fourth cylinders are covered with the new first and third cylinders that make complete complex cylinders.

Growing ends of complete complex cylinders attract each other and make new centrioles, which are attracted by the old centrioles in the centrosome bodies at the end of G2. New coiling phases start when uncoiling phases end also at the end of G2 and prophase, metaphase, and anaphase of mitosis. Coiling phases are shorter than uncoiling phases.

The protein spirals of the string

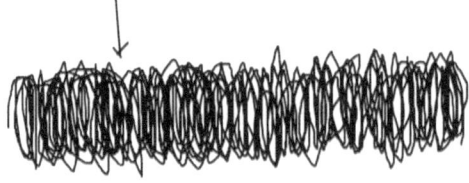

Protein spirals of the first and third cylinders are very densely coiled in the opposite directions at the end of anaphase.

Coiling of the first and third cylinders' strings of cylinders causes the cylinders to twist and become shorter and thicker. Coiling also causes a disappearance of centrosome bodies into centrioles and their separation into opposite poles of the cell. The shorter and thicker cylinders produce different centrioles, a disappearance of the nucleus membrane, and an appearance of mitotic spindles and chromosomes between centrioles during metaphase. The rationale for mitosis is that the different structures are connected by strings and cannot exist as the own entities.

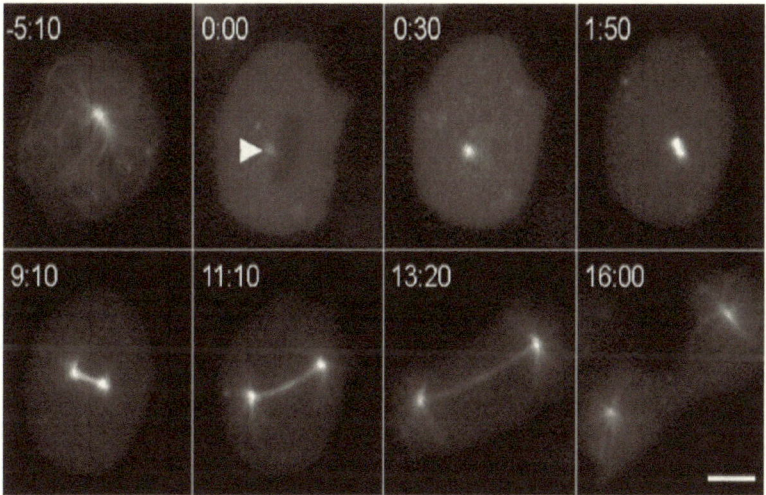

Movements of the two groups of complex cylinders to two opposite poles of a cell are due to reducing of lengths of the complex cylinders that transform into the strings with chromosomes in the mitotic spindle between centrioles. The sequence shows cell division in dictyostelium. I am very thankful to Dr. Michael Koonce for his pictures of the movements of centrioles to the opposite poles of the cell. The pictures enlighten the complex process.

New first and third cylinders increase the coiling of strings with their protein spirals during their growth around the second and fourth cylinders that attract them. Nucleic acid spirals of the new first and third cylinders decrease coiling during the growth of the strings due to their spending during their growth. The second and fourth cylinders of the cut complex cylinders decrease the coiling of strings during growth of strings of the first and third cylinders of the incomplete cylinders because they grow into new first and third cylinders. The second and fourth cylinders of the cut cylinders accumulate nucleic acids in strings because the new parts of protein and nucleic acid spirals get in at their ends.

Increasing coiling of strings of the first and third cylinders and decreasing coiling of nucleic acids in them and the decreasing coiling of strings of the second and fourth cylinders with increasing coiling of nucleic acids in them enable torsion, cutting, and division of the complex cylinders simultaneously and their growths (uncoiling) after mitosis. These phenomena can take place only if the coils of the strings attract each other during metabolism. Protein spirals of strings of complex cylinders have the characteristics of elastic iron springs as do protein spirals of the strings of filaments in prokaryotic cells. The characteristics of the elastic protein springs were explained during the explanation of the structures of prokaryotic cells.

Torsion and division of cylinders occur only if their strings move in opposite directions during metabolism and growth and if their strings attract each other. Structures of eukaryotic cells exist also according to the law of the unity and conflict of opposites.

In summary, complex cylinders transform into different structures during uncoiling and coiling phases. The incomplete cylinders become complete and produce nuclei with membranes, organelles, and centrosome bodies during the uncoiling phase. Cylinders with centrosome bodies transform into opposite centrioles, chromosomes with the disappearance of the nucleus membrane, and mitotic spindles during the coiling phase. Torsion reduces the length of cylinders producing different structures described in the different phases of mitosis. Coiling of the first and fourth cylinders creates torsion and division by the chewing gum phenomenon.

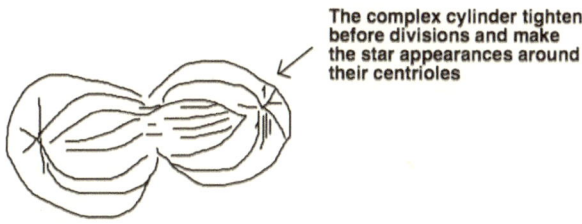

The complex cylinder tighten before divisions and make the star appearances around their centrioles

The phenomenon of chewing gum causes the shortening and torsions of the strings in complex cylinders producing their divisions. Complex cylinders look like rays around the centrioles due to their shortening and tightening. The cutting of the cell starts at the periphery because the longest complex cylinders at the periphery have the fastest coiling and divisions.

Mitosis is explained differently here than in the literature, which has no logical explanation for the different phases (appendix XXVIII) and their durations.

Mitosis has also several phases: interphase (includes G1, S, and G2), prophase, metaphase, anaphase, and telophase.

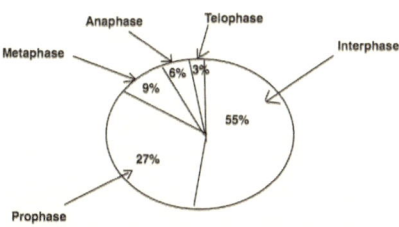

Time spent in each phase of mitosis

Interphase consists of G1, S, and G2; the facts from the literature are accepted and explained with the new approach. Uncoiling of cut and short complex cylinders occurs during telophase, G1, S, and a big part of the G2 phase. Cut, short, and densely coiled cylinders transform into long, incomplete cylinders that hit close daughter cells, bend, and come back along themselves due to their increased lengths and lack of space.

The nucleolus, nucleus, and cytoplasm with the immature organelles are products of the rolling of the bent and incomplete cylinders.

Duplication of chromosomes as separate bodies does not exist according to the new view. Duplication of nucleic acid and protein spirals exists in the strings of cylinders according to the new approach. S phase ends when maturation of mitochondria and endoplasmic reticulum with the other organelles ends.

Nuclei have a net of long, thin chromosomes called chromatin. The G2 phase starts when mitochondria and endoplasmic reticulum with the other organelles start to disappear and when nucleus chromatin starts to transform into shorter and thicker chromosomes. The G2 phase finishes when coiling and shortening of cylinders cause division of centrosome bodies into centrioles and when mitochondria disappear.

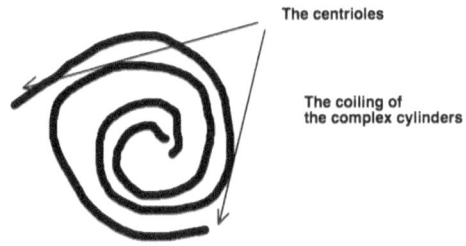

The centrioles

The coiling of
the complex cylinders

Coiling of complex cylinders causes opposite centrosomes.

Prophase starts when the nucleus membrane starts to disappear and chromatin transforms further into thin and long chromosomes due to the cylinders' coiling. Prophase finishes when the nucleus membrane disappears when centrioles have opposite positions with the appearance of mitotic spindles and when chromosomes mature; mature chromosomes do not get shorter and thicker. Strings of the cylinders make the mitotic spindles that connect centrioles and chromosomes due to unequal coiling of the cylinders.

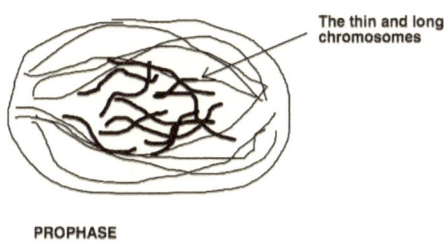

The thin and long chromosomes

PROPHASE

Metaphase starts when the thick and short chromosomes arrange in the middle cell.

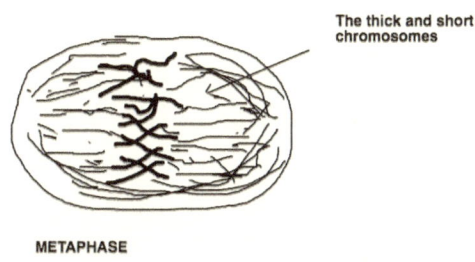

The thick and short chromosomes

METAPHASE

Mature chromosomes undergo constriction in centromeres due to continuous coiling of complex cylinders that produces torsion and narrowing. Constriction of a chromosome is also called a centromere also according to the new approach.

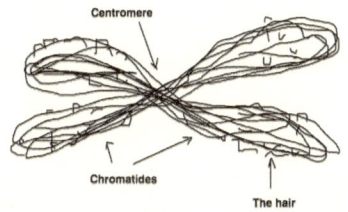

Centromere

Chromatides

The hair

Hairs of a pair of homologous chromosomes are products of densely coiled filaments (appendix XXIX).

Metaphase finishes and anaphase starts when pairs of cylinders with mature chromosomes start to divide and move to centrioles. Cell membranes and cytoplasm start to divide due to extreme coiling and torsion. Dividing cells stay intact due to attractions of the strings of the cylinders among themselves during mitosis. Anaphase finishes when cut chromosomes are very close to centrosomes and when they uncoil and transform into long, incomplete chromosomes.

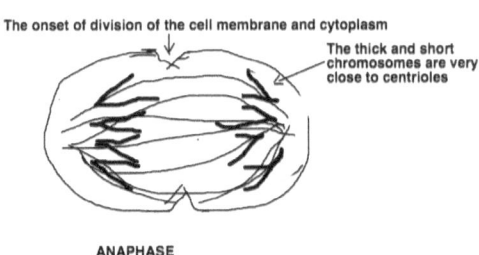

The onset of division of the cell membrane and cytoplasm

The thick and short chromosomes are very close to centrioles

ANAPHASE

Telophase starts when the constriction of the mother cell finishes and two daughter cells appear. The cut cylinders transform into long and incomplete cylinders that hit and bend in close daughter cells. They come back along the transformed complex cylinders and roll into nucleolus, nucleus, and cytoplasm. This hitting separates close daughter cells, which have diploid numbers of long cylinders and their chromosomes. The telophase lasts for a very short time.

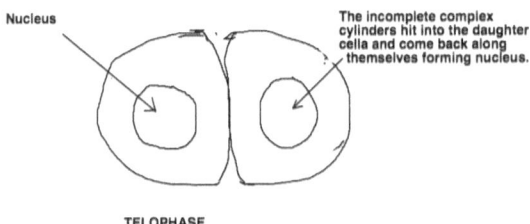

Nucleus

The incomplete complex cylinders hit into the daughter cells and come back along themselves forming nucleus.

TELOPHASE

In summary, a mother cell produces two daughter cells with diploid numbers of cylinders and their chromosomes during mitosis. Chromosomes are not duplicated; protein spirals of the first and third cylinders and nucleic acids in the second and fourth cylinders of the cylinders are duplicated. Coiling and uncoiling phases change alternatively during the lives of eukaryotic cells.

Hypothesis for Delayed Mitosis of Specific Cells

Many specialized cells (nerve cells, muscle cells, erythrocytes, plasmocytes, and others) have delayed mitosis despite the increased interchange of parts of protein and nucleic acid spirals and very active metabolism according to the new approach. Some cells (nerve and muscle cells) have delayed mitosis due to increased growth of cylinders without attraction into centrosome bodies necessary for mitosis. Centrioles of a centrosome body enable the formation of mitotic spindles and chromosomes; these cells have huge and long cells.

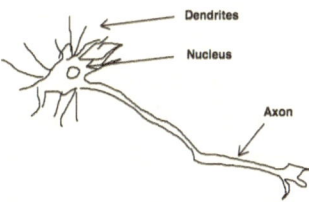

Axons and dendrites of nerve cells are associations of strings according to the new approach.

Some cells (erythrocytes, plasmocytes, and others) do not grow and divide because the protein spirals that move from straight parts to coiled parts of strings likely break into polypeptides at their coil parts or because the parts of the protein and the nucleic acid spirals are not attracted by the specific strings of their cylinders. The breaking of protein spirals at coil parts do not allow growth of complex cylinders and their cells, and the complex cylinders get even less. Therefore, some cells lose nuclei and have cytoplasm with small and poor organelles (erythrocytes).

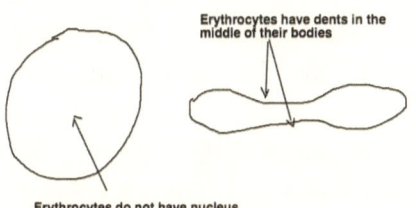

Erythrocytes do not have nucleus

213

Cells with long complex cylinders have increased life spans because they grow, whereas cells with short complex cylinders have decreased ones because their complex cylinders get less with reduced functions. Muscle cells associate in bigger cells with many nuclei because they have likely strong attractions among themselves.

Meiosis

Meiosis is a reduction of the diploid number of chromosomes in the predecessors of the sex cells that divide into daughter cells with a haploid number of chromosomes. The daughter cells mature into sex cells as gametes, pollen, or spores depending on eukaryotic species.

Opposite sex cells with a haploid number of chromosomes of the same species attract each other and fertilize into zygote cells with a diploid number of chromosomes. The zygote cells have mixed genes of the maternal and paternal sides. Mixed genes of zygote cells make different phenotypes of new organisms of the same species that develop from them.

Meiosis occurs in all sexually reproducing, single-celled, eukaryotic microorganisms (protists) and multicellular animals, plants, and fungi. These facts from the literature are accepted by the new approach that explains meiosis in another way. Meiosis is very similar to mitosis with additional phenomena according to the literature and the new approach. It occurs only in predecessors of sex cells very similar to stem cells according to the new approach because sex cells transform into zygote cells that must be similar to stem cells because they are born from them after mitosis.

Stem cells transform further into predecessors of sex cells and transitional cells that transform into specialized ones after mitosis. Zygote cells must have specific polypeptides that start reading the genetic plan of differentiation via stem cells into transitional cells and transitional ones into specialized cells. The predecessors of the sex cells have meiosis I and II also according to the new approach (appendix XXX).

Crossing over of complex cylinders and their chromosomes without their reductions occurs in meiosis I. The reduction of the diploid number of complex cylinders and chromosomes occurs at the end of meiosis II. Predecessors of sex cells also have coiling and uncoiling phases during meiosis I and II. Coiling occurs during a small part of G2 of interphase, prophase, metaphase, and anaphase. Uncoiling occurs during telophase, G1, S, and a big part of G2 of interphase.

Duplication of chromosomes does not exist in meiosis according to the new approach. Duplication of protein and nucleic acid spirals occurs in cylinders instead of duplication of the chromosomes according to the literature. Duplication of nucleic acid spirals occurs in the nongrowing parts of the strings in mitosis and meiosis I and II while duplication of protein spirals occurs in the growing parts of the strings.

Interchange of parts of protein and nucleic acid spirals must take place also among the strings of the pairs of cylinders of cells during meiosis I and II due to their growth. If cells reduce the diploid number of complex cylinders and their chromosomes into the haploid number during meiosis I, they cannot interchange parts of protein and nucleic acids during meiosis II. Therefore, the reduction of the diploid number of the cylinders and their chromosomes in cells occurs only at the end of meiosis II. Meiosis II produces sex cells with a haploid number of cylinders and their chromosomes.

Uncoiling occurs immediately after the division of the predecessor cells of the sex ones into daughter ones in meiosis I. The densely coiled strings of the first and third cylinders of the cut cylinders in the middle spread as springs into the new second and fourth cylinders, which increases the length of the transformed cylinders. These cylinders bend because they hit the daughter cells and come back along the transformed complex cylinders. The transformed and bent complex cylinders roll and make the nucleolus, nucleus, and cytoplasm of daughter cells in G1 of interphase I immediately after telophase. The ends of the long complex cylinders attract each other and make the new centriole, which is attracted by the old centriole making the centrosome body. The uncoiling phase is finished at the end of G2. The coiling phase also starts at the end of G2 and finishes with anaphase I. The crossing over occurs during G1 and prophase I of meiosis I. The new uncoiling phase starts in telophase I when meiosis II starts.

Crossing over occurs during G2 and prophase I of the cells during meiosis I according to the new approach while it occurs only in prophase I according to the literature. The hypothesis of the new approach is that very similar pairs of chromosomes and their cylinders attract one another in the tetrads. The hypothesis is also that pairs of the homologous chromosomes in the tetrads have very strong attractions that enable the crossing over only among homologous chromosomes and pairs of complex cylinders.

The strong attraction among homologous chromosomes and their movements by each other cause breaking at the similar weak points among the same polypeptides of the strings of the homologous complex cylinders and their chromosomes and their rotations in the tetrads. This breaking and rotation produce the exchange of similar parts of cylinders with their nucleic acids that rotated in the homologous chromosomes in the tetrads. The chromosomes must have similar weak points among the polypeptides that enable this according to the new approach. The similar parts attract the specific polypeptides of the homologous chromosomes in the tetrads after rotation and exchange (crossing over).

Tetrads enable the stable rotation and exchange of similar parts among the homologous chromosomes and unification among the same parts of the cylinders of the chromosomes during crossing over simultaneously. Tetrads are not duplications of the cylinders and their chromosomes according to the literature but attractions of similar pairs of homologous chromosomes into tetrads due to their specific attractions. The phenomenon of the tetrads according to the new approach is also explained in an example of the human precursor of the sex cells.

The human precursors of sex cells have forty-six complex cylinders and twenty-three pairs of homologous cylinders or twenty-three pairs of homologous chromosomes. The human precursors of the sex cells in meiosis I have eleven tetrads (11 x 4 = 44) with forty-four cylinders and a pair (2 x 1 = 2) with two homologous cylinders without a tetrad. They make altogether forty-six chromosomes with eleven tetrads, and two complex cylinders have duplicated nucleic acids without duplication of chromosomes according to the new approach. This hypothesis in the above paragraphs looks for proofs despite its logic.

| homologous complex cylinders A |
| homologous complex cylinders a |
| **homologous complex cylinders B** |
| homologous complex cylinders b |

Tetrads of homologous chromosomes are produced by attraction of similar pairs of cylinders during G2 and prophase I. One tetrad has two similar pairs of homologous chromosomes.

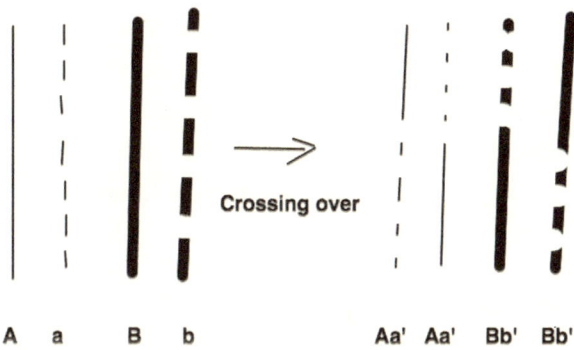

| A | a | B | b | | Aa' | Aa' | Bb' | Bb' |

Crossing over occurs among homologous chromosomes in tetrads at the cells during G2 and prophase I of meiosis I according to the new approach. Aa' and Bb' are mixed chromosomes after the crossing over.

Prophase I of the cells starts when the nucleus membrane starts to disappear and chromatin appears also according to the new approach. Chromatin transforms into chromosomes united in tetrads. Prophase I finishes when nucleus membranes do not exist and when centrioles come to opposite positions of the cells. Mitotic spindles exist between centrioles (centrosomes) and are connected to the tetrads, which are still united in this phase. Coiling of cylinders causes their separation and disappearance of the tetrads.

The coiling of the complex cylinders inside the tetrads causes separation of the complex cylinders and disappearance of the tetrads.

Metaphase I of cells starts when crossing over among chromosome tetrads is rare and when chromosomes tetrads start to disappear. The mitotic strings of the cylinders connected with centrioles connect separated chromosomes and tetrads. Constriction appears at chromosomes due to continuous coiling of their cylinders. Constriction at a chromosome is known as a centromere according to the literature and the new approach.

Coiling of cylinders and chromosomes promotes their separation. The disappearance of tetrads is due to densely coiled cylinders that do not allow full attraction among the cylinders and their chromosomes. The tetrads disappear and the crossing over stops because further coiling causes separation of the tetrads necessary for this phenomenon. Metaphase I finishes when cylinders and their chromosomes come to mid cell, the equator of the mitotic spindle. Metaphase I and other phases result from continuous coiling of cylinders.

Anaphase I of cells starts when chromosomes start to cut due to coiling and ends when the cut chromosomes travel to centrioles due to the densely coiled cylinders that pull them in opposite directions due to their transformations into incomplete cylinders and when cell membranes start to constrict and divide. Anaphase I is very short.

Telophase I starts when cylinders are cut and ends when two daughter cells are formed. The cells have a diploid number of mixed cylinders and their chromosomes after meiosis I. They interchange parts of protein and nucleic acids among the homologous complex cylinders and chromosomes.

The strings of the second and fourth cylinders of the cut cylinders in the middle grow around the new second and fourth cylinders of the new and incomplete cylinders as new first and third cylinders. This occurs in G1 and S of interphase II of meiosis II. Uncoiling ends when the new second and fourth cylinders are covered with the new first and third cylinders of complete cylinders. This occurs at the end of G2 of interphase of meiosis II.

New coiling phases start when uncoiling phases end; this also occurs at the end of G2 of interphase II, prophase II, metaphase II, and anaphase II. The cylinders do not attract each other in tetrads during prophase II. Coiling phases are much shorter than uncoiling phases. Half of the complex cylinders and their chromosomes are disintegrated in anaphase II before telophase II. The predecessors of sex cells, now sex cells, have a haploid number of cylinders and chromosomes after meiosis II. They do not interchange protein and nucleic acids among the cylinder pairs because they have a haploid number of cylinders and chromosomes. If they do not exchange protein and nucleic acids, they cannot grow. They metabolize as other cells, and if they do not fertilize soon in zygote cells, they die. Their life span is very short.

The daughter cells get into meiosis II, which has the same phases without tetrads and crossing over. Their homologous and mixed complex cylinders interchange protein and nucleic acids. The similar pairs of the homologous chromosomes do not attract each other in prophase II because the homologous cylinders of daughter cells are different from their mother cells. The interchange enables growths and coiling of their mixed cylinders and the described phases. Half of mixed complex cylinders cannot withstand further coiling and disintegrate into pieces during metaphase and anaphase producing a haploid number of cylinders and chromosomes. The precursors of the sex cells in meiosis II have similar weak points among their polypeptides.

Disintegrated complex cylinders take positions outside complex cylinders during reduction of chromosomes and their complex cylinders. These pieces will be parts of the protective layers at oocytes or will be out of spermatozoids.

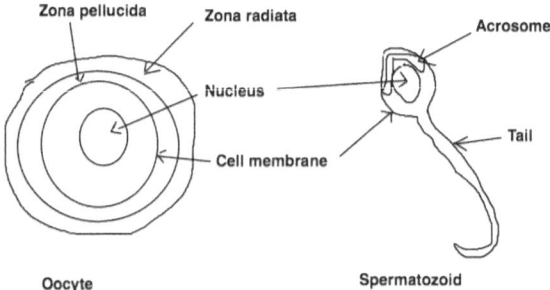

Oocyte · Spermatozoid

Oocyte cell with the haploid number of the complex cylinders has remains of broken complex cylinders.

A zygote cell is produced by unification of sex cells with a haploid number of cylinders due to their attractions.

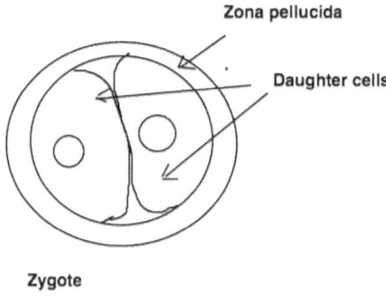

Zygote

Zygote cells divide by mitosis producing stem cells.

The homologous long cylinders of sex cells attract each other and produce a diploid number of cylinders in the zygote cell. This is possible only if the cytoplasm has cylinders that build nuclei and other organelles as a united system. Zygote cells transform into stem cells after mitosis. Stem cells start differentiating and produce cells that transform into transitional cells and predecessors of sex cells. The transitional cells transform further into specialized ones.

Chromosomal crossing over and fertilization of sex cells enable the double mixing of maternal and paternal genes in eukaryotic organisms. This crossing over and fertilization are continuous experiments of nature that produce combinations of similar characteristics in offspring. Environmental conditions choose those qualities; some combinations of genes and their phenotypes of the same eukaryotic species are excluded due to natural selection under the same environmental conditions.

Numerical and structural chromosomal abnormalities appear during crossing over in meiosis I or during reductions of cylinders in meiosis II. The long and mixed complex cylinders can or cannot withstand coiling during metaphase II.

The existence of an extra chromosome at a diploid number of chromosomes—trisomy—clearly shows that some cylinders are not broken, deactivated, disassembled, and utilized during meiosis II. The existence of monosomy (diploid chromosomes missing a chromosome) also shows that one extra complex cylinder with chromosomes is broken, deactivated, disassembled, and utilized. Some aneuploidy in humans can be tolerated with severe developmental or asymptomatic disorders:

- Down syndrome—trisomy of chromosome 21 (physical growth delays, characteristic facial features, and mild to moderate intellectual disabilities)
- Patau syndrome—trisomy of chromosome 13 (cleft palate, small eyes, reduced distance between eyes, absence of one or both eyes, smaller head size)
- Edward syndrome—trisomy of chromosome 18 (small head, mouth and jaw; prominent back of head, cleft lip or palate, hands clinched, clubfeet, webbed or fused toes)
- Klinefelter syndrome—extra X chromosomes in males (e.g., XXY, XXXY, XXXXY, etc.; small testes, female pubic hair, breast development, narrow shoulders, poor beard growth, etc.)
- Turner syndrome—lacking one X chromosome in females (i.e., X0; short stature, characteristic facial features, brown spots, poor breast development, small finger nails, no menstruation, rudimentary ovaries, constriction of aorta)
- Triple X syndrome—an extra X chromosome in females (moon face, thinning of scalp hair, acne, obesity, muscle wasting, bruising, constriction of aorta, hypertension, osteoporosis, mental instability)

- XYY syndrome—an extra Y chromosome in males (taller, acne, and learning disorders)

Differentiation of Stem Cells

An understanding of stem cell differentiation into transitional and then specific ones is necessary to understand the transformation of normal cells into malignant ones according to the new approach. The important facts are these.

- Cells have a diploid number of chromosomes half of which are inherited from each parent.
- Two similar chromosomes from the male and female parents are called a homologous pair; they have the same genes at the same loci. The same genes of the same loci of homologous chromosomes are different. The gene of the same loci has two alleles.

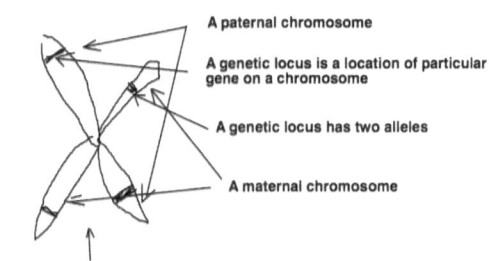

A paternal chromosome

A genetic locus is a location of particular gene on a chromosome

A genetic locus has two alleles

A maternal chromosome

A pair of homologous chromosomes

- Homologous chromosomes of cells are different because they have different genes of the same loci that produce different proteins of the same functions.
- A chromosome has four different ds DNA according to the new approach. A chromosome is a part of the complex cylinder that has four strings with four ds DNA. Each string has ds DNA that is active during biosynthesis of polypeptides and nucleic acids.
- Cylinders consist of four cylinders on their ends unified in the two in the middle.

- Cylinders are made of four strings that bring nucleic acids.
- Nucleic acids from the strings that build the first and second cylinders of the same pair are very similar though they build two different strings. The strings of the first and second cylinders are synthetic and decomposition. Nucleic acids from the strings that build the third and fourth cylinders of the same pair are also very similar though they build two different strings. The strings of the third and fourth cylinders are also synthetic and decomposition.
- Nucleic acids of the first and second cylinders are different from those of the third and fourth cylinders.
- Complex cylinders of cells are organized in pairs to enable interchange of protein spirals and nucleic acids. The first and second and the third and fourth cylinders interchange nucleic and protein spirals among themselves.
- Complex cylinders of the pair are very similar regarding proteins, nucleic acids, and lengths. The pairs of complex cylinders of a cell are different regarding quantities and qualities of protein and nucleic acid spirals including their lengths.
- Eukaryotic cells of different species have different numbers of chromosomes that are parts of the complex cylinders. Specific pairs of specific complex cylinders interchange specific parts of protein and nucleic acid spirals. See examples of different numbers of chromosomes of the different species in the next table.

Organism	Species	Diploid number of chromosomes
dog	*Canis familiaris*	78
donkey	*Equus asinus*	62
frog	*Rana pipiens*	26
fruit fly	*Drosophila melanogaster*	8
avocado	*Perse Americana*	24
human	*Homo sapiens*	46

- Human eukaryotic cells have twenty-three pairs of chromosomes and complex cylinders.

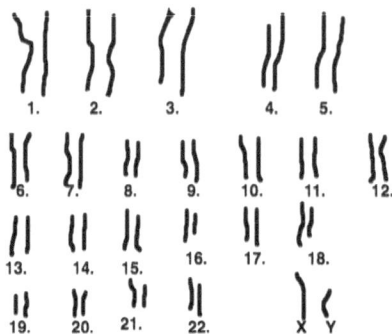

- Different genes along nucleic acids of the strings are organized in groups that repeat along nucleic acids in homologous chromosomes;

A	B	C	A	B	C

- Any different group has sixteen genes with two subgroups, which have eight different genes. One subgroup synthesizes polypeptides for synthetic protein spirals and another synthesizes polypeptides for decomposition protein spirals.

1	1'	2	2'	3	3'	4	4'	5	5'	6	6'	7	7'	8	8'

One group of genes consists of two subgroups of the genes that permeate each other. Each subgroup has eight genes. The diagram shows only a group of genes that appear among ds DNA of the first, second cylinders, third, and fourth cylinder.

- Eight genes of both subgroups synthesize eight polypeptides that synthesize N-acetyl-D-glucosamine and glucuronic acid. N-acetyl-D-glucosamine and glucuronic acid repeat alternatively in huge molecules of hyaluronic acids, which can be synthesized only if the specific groups of polypeptides and coenzymes repeat along the protein spirals. Groups of polypeptides with coenzymes that repeat along the protein spirals can be synthesized only if the gene groups exist along nucleic acids that can synthesize them.

- Two subgroups of genes of the same group permeate each other due to the polypeptide and nucleic acid biosynthesis according to the new approach. A subgroup of genes attracts polypeptides and frees the other subgroup of the genes during the synthesis.
- Different quantities of groups of genes produce different quantities of polypeptides for different complex cylinders that form different volumes, shapes, and qualities of the cells.
- Different genes of the same group that repeat along ds DNA are very similar but not the same. Very similar genes in the same site (loci) of homologous chromosomes are known as alleles.

There are three scenarios of how different groups of different genes exist in complex cylinders and synthesize a specific cell. The first scenario is that pairs of complex cylinders of a cell have active groups of "similar" genes that determine their characteristics. The facts of different pairs of homologous chromosomes with active different genes are against this scenario.

The second scenario is that pairs of the cylinders of a specific cell have different groups of genes that determine its characteristics and they are active only for the specific cells and not for other specific cells. This scenario is possible because different pairs of different homologous chromosomes exist in different groups of genes. The facts of the same proto-oncogenes in the different cells of an organism are against this scenario. Other words the same proto-oncogenes or the same genes appear in the different cells.

The third scenario is a modification of the second: the different cells of an organism have different pairs of homologous complex cylinders with active groups of genes that determine their characteristics. The different cells of an organism have also some the same groups of genes active in some different cells. The third scenario does not contradict the facts of the different pairs of the homologous chromosomes produced by the different genes of a cell and appearance of the same proto-oncogenes or the same genes in the different cells of an organism.

Proto-oncogenes are genes that mutate into oncogenes, genes that produce oncopolypeptides, which cause malign alterations of cells by unknown mechanisms according to the literature. The proto-oncogenes belong to different groups of genes active in some different cells according to the new approach. A good example is P53, which is present in different cells that produce brain tumors, sarcomas, breast cancers, and leukemia, which are born from different cells. Many other examples exist. This is possible only if the same proto-oncogenes belong to different groups of similar genes that build some the same complex cylinders in different cells. See the table below with the different proto-oncogenes.[12]

Proto-oncogenes	Specific cells that bring active proto-oncogenes	Tumors
STK11 = tumor suppressor	unknown cells	hyperpigmentation, multiple hamartomatous polyps, colorectal, breast and ovarian cancers
MSH2 = tumor suppressor	unknown cells	colorectal cancer
MLH1 = tumor suppressor	unknown cells	colorectal cancer
VHL = tumor suppressor	unknown cells	renal cancers, hemangioblastomas, pheochromocytoma
CDKN2A = tumor suppressor	unknown cells	melanoma, pancreatic cancer, others
PTCH = tumor suppressor	unknown cells	basal cell skin cancer
MEN1 = tumor suppressor	unknown cells	parathyroid and pituitary adenomas, islet cell tumors, carcinoid
MEN2, also known as RET	unknown cells	medullary thyroid cancer, type 2A pheochromocytoma, mucosal hartoma
BWS	unknown cells	genomic imprinting disorder resulting in Wilms tumor, adrenocortical cancer, hepatoblastoma
MET	unknown cells	renal papillary cancer
PTEN	unknown cells	breast cancer, thyroid cancer, head and neck squamous carcinomas

HPC1 and PRCA1	unknown cells	prostate cancer
ATM	unknown cells	lymphoma, cerebellar ataxia, immunodeficiency
XPA, ERCC3 (XPB), XPC, ERCC2 (XPD), DDB2 (XPE), ERCC4 (XPF), ERCC5 (XPGC), and variant XP (XPV, POLH gene)	unknown cells	skin cancer
FANCA, B, C, BRCA2 (D1), D2, E, F, G, I, BRIP1 (J), L, M, PALB2 (N), RAD51C (O), SLX4 (P), ERCC4 (Q), RAD51 (R), BRCA1 (S), UBE2T (T)	unknown cells	acute myeloid leukemia (AML), pancytopenia, chromosomal instability
P53 = tumor suppressor	unknown cells	brain tumors, sarcomas, leukemia, breast cancer
RB_1 = tumor suppressor	unknown cells	retinoblastoma, osteogenic sarcoma
WT1 = tumor suppressor	unknown cells	pediatric kidney cancer
NF1 = tumor suppressor	unknown cells	neurofibromas, sarcomas, gliomas
NF2 = tumor suppressor	unknown cells	Schwann cell tumors, astrocytomas, meningiomas, ependynomas
APC = tumor suppressor	unknown cells	colon cancer
TSC1= tumor suppressor	unknown cells	seizures, mental retardation, facial angiofibromas
TSC2 = tumor suppressor	unknown cells	benign growths (hamartomas) in many tissues, astrocytomas, rhabdomyosarcomas
DPC4 = tumor suppressor	unknown cells	pancreatic carcinoma, colon cancer
DCC = tumor suppressor	unknown cells	colorectal cancer
BRCA1 = tumor suppressor	unknown cells	breast and ovarian cancer
BRCA2 = tumor suppressor	unknown cells	breast and ovarian cancer

Different groups of gene pairs of complex cylinders are organized in bigger groups that repeat along their ds DNA according to the plan of differentiation of stem cells into transitional ones and then specialized ones. The next scheme shows the transformation of stem cells into specialized cells without transitional ones. This scenario without transitional cells does not exist.

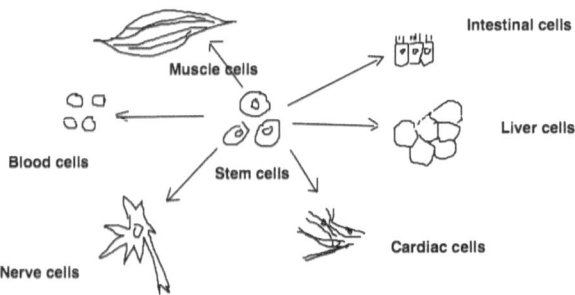

Cell differentiation without transitional cells does not occur.

According to the new approach, differentiation of stem cells into transitional and then different ones are explained by different quantities and qualities of metabolic products that appear in stem and transitional cells. Different metabolic products change attraction of polypeptides for genes of the new subgroups of the new different groups according to the plan of differentiation.

The new subgroups of the new different groups of genes produce polypeptides for new specific cells according to the plan of differentiation. This approach must be proven by experiments. There are the three important points regarding differentiation of the stem cells via transitional ones into specialized ones according to the new approach.

1. An organism's cells have different volumes, shapes, and qualities and the same number of pairs of complex cylinders. The cells have different quantities of polypeptides, protein spirals, strings, and different sizes of complex cylinders that build them. The gene groups along ds DNA of the strings of pairs of complex cylinders produce different qualities and quantities of polypeptides, protein spirals, strings, and complex cylinders of different cells. These gene groups repeat along nucleic acids, which are almost the same for an organism's cells. They are not exactly the same in all different cells of an organism because

mutations appear during mitosis of stem, transitional, and specific cells.

From ectoderm, endoderm and mesoderm as transitional cells develop transitional cells with active specific groups of genes. From transitional cells of the ectoderm develop the specific cells with active specific groups of the genes that make integumentary systems (fifteen keratinizing epithelial cells), wet stratified barrier epithelial cells (three), nervous system (twenty sensory transducer cells), autonomic neuron cells (three basic cells), sense organ and peripheral neuron supporting cells (twelve cells), central nervous system neurons and glial cells (four) and lens cells (two).

Specific cells with active specific groups of genes that make exocrine secretory epithelial cells (fourteen) and hormone-secreting cells (thirty-one) develop from specific transitional cells of the endoderm.

Cells with active groups of genes that make active metabolism and storage (four), kidney cells (seventeen), extracellular matrix cells (twenty-one), contractile cells (twelve), blood and immune system cells (twenty-one), germ cells (five), nurse cells (three) and interstitial cells (one) develop from specific transitional cells of the mesoderm.

2. Stem cells differentiate into transitional ones by attraction of their polypeptides for genes of transitional cells. Transitional cells become specialized by attracting polypeptides of the specialized cells. This attraction occurs according to the plan of differentiation under metabolic conditions produced by cells. The polypeptides of specific cells are products of genes of the groups of genes along nucleic acids.

Subgroups of genes permeate other subgroups of genes of a group due to polypeptide and nucleic acid synthesis. Subgroups of genes attract polypeptides and free the other subgroup of genes during polypeptide synthesis.

Attraction of polypeptides of stem cells for genes of groups of transitional cells occurs most likely due to concentrations of metabolites produced by the cells; that changes the attraction of polypeptides for transitional cells. This hypothesis must be proven. This attraction occurs most likely due to metabolites produced by transitional cells. The concentration of metabolites produced by transitional cells changes attraction of specific polypeptides of transitional cells for genes of specific groups of specialized cells. This hypothesis must be also proven.

This attraction of polypeptides occurs despite bigger quantities of groups of genes of transitional cells than quantities of polypeptides of stem cells produced by smaller quantities of groups of stem cell genes.

Attraction jumps of polypeptides of transitional cells for gene groups of specialized cells also takes place despite bigger quantities of gene groups of specialized cells than quantities of polypeptides of transitional cells produced by smaller quantities of groups of specific genes of transitional cells because polypeptides move by spiral movement with close polypeptide spirals for interaction with positive DNA strands. Concentration of polypeptides in the protein spirals even smaller than genes attract them and free gene groups during biosynthesis of polypeptides and nucleic acids (appendix XXXI).

Quantities of gene groups for cells in a genome of a cell determine quantities of polypeptides, sizes of protein spirals, and sizes of complex cylinders. These determine qualities, shapes, sizes, metabolism, bioelectricity, frequency of depolarization, and frequencies of mitosis of cells. Cells' metabolism, movement, bioelectricity, frequency of depolarization, and frequency of mitosis determine their quantities and organization in tissues and organs and different shapes of complex organisms.

The different groups of genes are organized in the biggest groups which repeat along nucleic acids of the strings of complex cylinders.

I	II	III	I	II	III	I	II	III
I	II	III	I	II	III	I	II	III

The above scheme shows the biggest groups of genes that also repeat along nucleic acids of the strings of the homologous complex cylinders.

The biggest groups contain different and same groups of genes and therefore they are not the same. The groups of genes of the biggest groups transform into different groups of polypeptides according to the strict order of differentiation of stem and transitional cells.

Stem cells have shorter complex cylinders, strings, protein spirals, and polypeptides than other cells do. The reverse is true for specialized cells, while transitional cells are between the first two in this matter. Many specialized cells are bigger than transitional and stem cells because they have longer protein spirals with longer polypeptides that make longer strings and complex cylinders. Stem cells always produce transitional ones, which produce specialized ones, not the opposite. It means shorter polypeptides of protein spirals of stem or transitional cells always attract longer genes according to the plan of differentiation. An exception appears during transformation of normal cells into malignant ones; that will be explained during the discussion of malignant cells.

Different multicellular eukaryotic organisms have different cells with the same nucleic acids that were born from different stem cells by their differentiation. Similar species have very similar differentiations which are product of their close evolutionary connections. Stem cells are born by the unification of the sex cells with a haploid number of chromosomes and their complex cylinders. The stem cells divide by mitosis into the transitional cells that make morula, blastula, and gastrula forms.

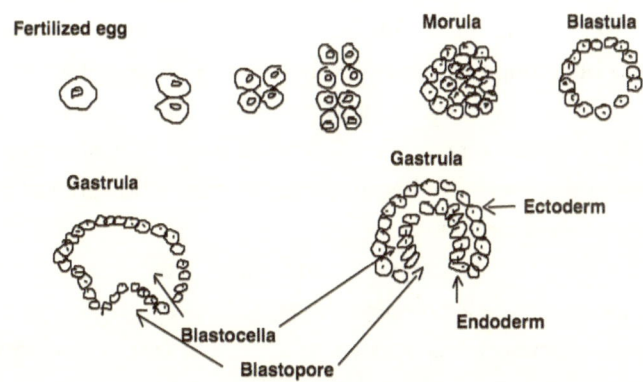

Blastula forms transform into gastrula by increased mitosis of transitional cells. Gastrula forms have an empty space (blastocella) and transitional cells that differentiate into ectoderm, mesoderm, and endoderm cells that are also transitional cells. Ectoderm, mesoderm, and endoderm cells of a gastrula differentiate into precursors of specific cells according to the specific plan explained by the below scheme from the literature.

Endoderm	lung cells, thyroid cells, digestive cells, and others
Mesoderm	cardiac muscle cells, skeletal muscle cells, tubule cells of the kidney, smooth muscle cells, red blood cells, and others.
Ectoderm	skin cells, pigment cells, nerve cells, and others

3. The same cells of an organism group in tissues likely due to their electrical attractions. They attract each other by the cell membranes which have opposite electric charges and similar frequencies of polarization and depolarization. The tissues of the specialized cells attract tissues of other specialized cells into organs for the same reasons and duo to interchange of metabolites. Organization of tissues and organs in organisms are products of their movements and interchanges of metabolites during their continuous fighting with environmental conditions and their evolution.

Groups of genes have strict organization in the biggest groups of genes along different ds DNA of pairs and numbers of complex cylinders. The groups of genes likely produce different shapes, sizes, and qualities of tissues, organs, and organisms. The groups of genes at the biggest groups along nucleic acids of different cells of organisms are not known in the literature because their nucleic acids are not explored according to this approach.

Species with evolutionary connections have similar orders of attraction among specific polypeptides and genes of stem and transitional cells during embryogenesis because of similar ancestry. Embryogenesis of different animal species shows that different animal organisms have similar plans of differentiation of stem, transitional, and specialized cells. The scheme shows their similarities during their development.

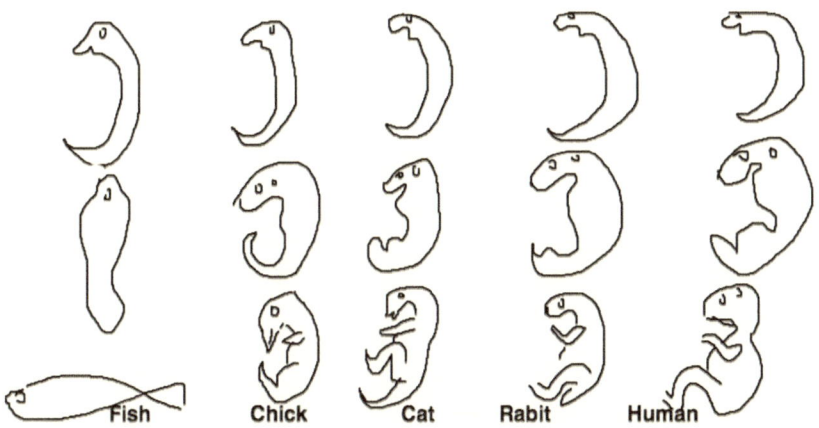

Embryogenesis of different animal species is very similar.

CHAPTER 19

HYPOTHETICAL EVOLUTION OF EUKARYOTIC CELLS AND UNIVERSAL LAWS OF NATURE

Eukaryotic cells most likely had similar predecessors and development as prokaryotic cells up to the pairs of the archaic strings (appendix XXXII). After the pair of archaic strings, the predecessors of prokaryotic and eukaryotic cells developed differently. Predecessors of eukaryotic cells appeared when archaic, very long strings transformed into structures with two cylinders at the ends and one in the middle. (See the hypothetical evolution of prokaryotic cells in chapter 14 before the hypothetical evolution of eukaryotic cells.) The hypothetical archaic forms of predecessors of eukaryotic cells are these:

- archaic polypeptides
- archaic protein spirals
- archaic strings
- pairs of complex cylinders with two cylinders at opposite sides and a cylinder in the middle
- pairs of archaic complex cylinders with four cylinders on opposite sides and two in the middle
- archaic eukaryotic cells
- eukaryotic cells of archaic colonies
- eukaryotic cells of simple archaic organisms
- eukaryotic cells of complex archaic organisms

Archaic polypeptides, protein spirals, and strings were predecessors of prokaryotic and eukaryotic cells as discussed in chapter 14. The difference between archaic polypeptides of prokaryotic and eukaryotic cells is that the polypeptides of the predecessors of eukaryotic cells were longer than the polypeptides of the predecessors of prokaryotic cells. The longer archaic polypeptides had more coenzymes and nonprotein hormones that brought more and greater quantities of ions. The greater quantities of ions enabled greater attraction among archaic polypeptides, which enabled longer archaic strings of predecessors of eukaryotic cells than archaic strings of predecessors of prokaryotic cells. The longer strings enabled specific structures of archaic complex cylinders that had been present at the predecessors of eukaryotic cells.

Pairs of Archaic, Long, Complex Cylinders with Two Cylinders on Opposite Sides and One in the Middle

Archaic, long synthetic and decomposition strings attracted and twisted around each other forming two cylinders on opposite sides and a cylinder in the middle.

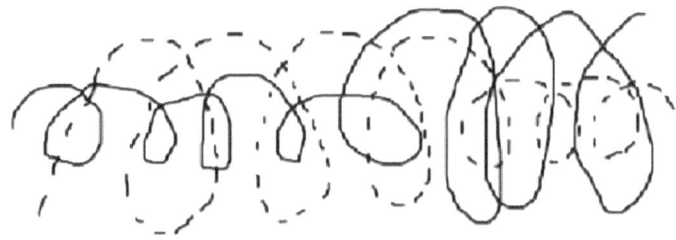

Outside archaic cylinders had been the growing parts of archaic decomposition strings on a side and growing parts of the archaic synthetic string on the other side. Inside archaic cylinders had been the nongrowing parts of archaic synthetic strings on the side of the outside archaic cylinders of the growing parts of the archaic decomposition string and nongrowing parts of archaic synthetic strings on the other side.

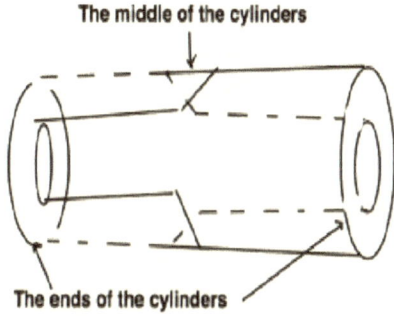

The drawing presents the two separate cylinders in the ends and unified cylinders in the middle.

Archaic structures with two cylinders at the ends and one in the middle lived in pairs due to interchange of parts of protein spirals and nucleic acids. The pairs attracted each other where decomposition cylinders attracted synthetic cylinders and vice versa. These lived in colonies due to better survival rates under different environmental conditions. They had different diameters and lengths. These structures negated pairs of archaic and very long strings. Unknown environmental conditions caused their evolution.

Pairs of Archaic Complex Cylinders with Four Cylinders on Opposite Sides and Two in the Middle

Structures of greater diameters swallowed smaller structures due to their metabolic needs and formed cylinders with four cylinders at the ends and two in the middle. Later cylinders lived in pairs due to attraction and interchange of protein spirals and nucleic acids. They also lived in colonies for better survival.

Archaic complex cylinders and their associations were predecessors of archaic eukaryotic cells. Their divisions had been similar to divisions of the archaic filaments.

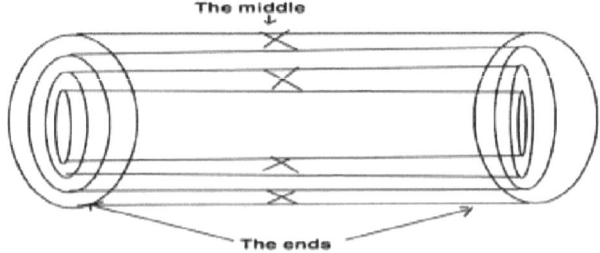

Similar pairs of archaic complex cylinders with weaker attraction united in weaker associations but did not develop into archaic eukaryotic cells because the attraction forces were too small.

Archaic complex cylinders with strong attraction forces united in stronger associations; these archaic eukaryotic cells had a stable number of long complex cylinders and developed into archaic eukaryotic cells because the pairs had stronger forces of attraction.

Archaic Eukaryotic Cells

Unknown environmental conditions forced pairs of cylinders with strong attraction forces into archaic long eukaryotic cells without nuclei, the Golgi apparatus, or centrosome, but they most likely had endoplasmic reticulum, ribosomes, cell membranes, mitochondria, and chromosomes before their divisions. These cells most likely had long bodies of different diameters and sizes depending on the number of pairs of long complex cylinders. They interchanged parts of protein spirals with nucleic acids outside archaic cells. They divided similar to divisions of archaic filaments, which had the same divisions as contemporary filaments of prokaryotic cells.

Harsh environments reduced the long bodies and exposed surfaces of archaic long cells by spiraling or the U bending and rolling into balls to reduce their exposed surfaces. They did not have nuclei but did have the same organelles as the archaic long cells did. They interchanged protein spirals with nucleic acids outside archaic cells at opposite poles.

Spiral bodies of archaic eukaryotic cells reduced their lengths and thus their exposed surfaces.

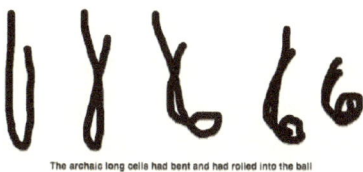

The archaic long cells had bent and had rolled into the ball

U-shaped bodies reduced their lengths and exposed surfaces to environments.

The U body rolled into a nucleolus, a nucleus with the chromosomal material (the chromosomes had appeared during mitosis), nucleus membrane, and the Golgi apparatus; they rolled into cytoplasm with organelles (endoplasmic reticulum, mitochondria, and predecessors of centrioles) and cell membranes.

The nucleus rolled the rest of the body of the archaic cells in the cytoplasm with organelles like a snowball.

Complex cylinders of archaic cells had cones on their ends as predecessors of centrioles due to attraction of their strings. They had interchanged protein and nucleic spirals outside archaic cells at opposite poles. Archaic cells had not had centrioles that appeared in contemporary eukaryotic cells; they had organelles that differed from contemporary eukaryotic cells except lysosomes due to the outside interchange of protein and nucleic acid spirals. Lysosomes are the cisterns of the endoplasmic reticulum that are products of the interchange of protein and nucleic acid spirals in endoplasmic reticula. They lost protein and nucleic acid spirals under different environmental conditions, and that hindered their ability to metabolize and grow.

Strings of archaic eukaryotic cells that had bent stopped losing protein and nucleic acid spirals because they interchanged them in the cytoplasm. The strings' change of directions transformed the cones in the centrioles and produced accumulations of protein spirals among the complex cylinders, which produced lysosomes. Archaic cells thus transformed into archaic eukaryotic cells very similar to contemporary ones. Unknown environmental conditions transformed archaic eukaryotic cells into contemporary ones.

Eukaryotic Cells of Archaic Colonies

Environmental factors forced eukaryotic cells with inner nuclei to associate in colonies. Nucleic acids of dead archaic cells were attracted and fused by nucleic acids of similar complex cylinders of other archaic cells. The enriched nucleic acids of the surviving archaic cells enabled their survival under different environmental conditions by changing their phenotypes after mitosis. Archaic cells with enriched nucleic acids developed into eukaryotic cells of simple organisms.

Some archaic cells with enriched nucleic acids did not survive, and their enriched nucleic acids fused with enriched nucleic acids of other archaic cells. This process of evolution produced colonies with very enriched nucleic acids that transformed into complex organisms.

Eukaryotic Cells of Simple Archaic Organisms

Unknown environmental conditions thus changed archaic colonies with different cells into simple archaic organisms (appendix XXXIII) with different eukaryotic cells. The eukaryotic cells had not lost their ability to change into phenotypes according to environmental conditions because they had nucleic acids with groups of different genes but not meiosis and archaic sex cells that produced archaic zygote cells. Meiosis had appeared at complex archaic organisms.

Eukaryotic Cells and Complex Archaic Organisms

Eukaryotic cells with very enriched nucleic acid lost the ability to transform into phenotypes for that reason. They developed cells organized in tissues of complex organs. They also developed reproductive organs that produced sex cells with a haploid number of chromosomes. That enabled wider combinations of genes with different characteristics for better survival rates. The phenomenon of crossing over that appeared during meiosis enabled different combinations of groups with similar genes.

Archaic zygote cells had a diploid number of chromosomes and produced archaic stem cells and predecessors of archaic sex cells. Archaic stem cells differentiated into transitional cells that differentiated into specific cells of different tissues of complex organs also according to the specific order.

Universal Laws of Nature and Eukaryotic Cells

Law of Unity and Conflict of Opposites

Eukaryotic cells of different species have different numbers of pairs of complex cylinders and their chromosomes that make the unity by attracting each other and conflict of opposites by their qualities. Pairs of complex cylinders attracting each other build different structures and organelles and different forms and sizes at different eukaryotic cells. The pair of complex cylinders of eukaryotic cells has similar complex cylinders and almost the same nucleic acids. The similar complex cylinders with almost same nucleic acids of the pair interchange parts of protein and nucleic acid spirals among themselves making the unity and conflict of opposites.

The four strings of complex cylinders build the growing coils on opposite sides of the first and third cylinders and the nongrowing coils at opposite sides of the second and fourth cylinders making the unity and conflict of opposites. The four strings of the four cylinders of the complex cylinders unite into the outside and the inside cylinder in its middle making the unity and conflict of opposites.

The four strings of the four cylinders from both sides are unified into the two cylinders in the middle of the complex cylinder.

Complex cylinders consist of two decomposition and two synthetic strings that make the unity and conflict of opposites. The different decomposition and synthetic strings have different decomposition and synthetic protein spirals that move in opposite directions by each other making the unity and conflict of opposites.

Decomposition strings produce decomposition spirals in synthetic spirals that move from the second to the first cylinder or from the fourth to the third one on opposite sides of the cylinders. Decomposition strings have spirals with tRNA that move from the first cylinder to the second or from the third to the fourth opposite of their synthetic protein spirals that bring mRNA. The spirals of the decomposition string make the unity and the conflict of opposites. Synthetic strings produce synthetic spirals in decomposition spirals that move similarly but opposite that of decomposition protein spirals that bring mRNA. The spirals of the synthetic string make also the unity and the conflict of opposites.

Complex cylinder 1

Growing part of synthetic string of the first (third) cylinder	Growing part of decomposition string of the first (third) cylinder
Nongrowing part of decomposition string of the second (fourth) cylinder	Nongrowing part of synthetic string of the second (fourth) cylinder

Complex cylinder 2

Growing part of synthetic string of the first (third) cylinder	Growing part of decomposition string of the first (third) cylinder
Nongrowing part of decomposition string of the second (fourth) cylinder	Nongrowing part of synthetic string of the second (fourth) cylinder

The scheme shows the specific positions of the growing and the nongrowing parts of synthetic and decomposition strings of the first and second or the third and fourth cylinder of the complex cylinder.

Synthetic and the decomposition strings of the first and second cylinders and those of the third and fourth cylinders have almost the same nucleic acids that make the unity and conflict of opposites by producing the opposite protein spirals.

Nucleic acids of the first and second cylinders are different from those of the third and fourth. They make the unity and conflict of opposites that produce opposite protein spirals in the cylinders. The first and second cylinders and the third and fourth cylinders interchange parts of protein and nucleic acid spirals making the unity and conflict of opposites. See the below scheme that helps to understand the above sentences.

Synthetic string of complex cylinder 1

Synthetic protein spiral >>>	Synthetic protein spiral >>>
<<< New synthetic protein spiral	<<< Decomposition protein spiral

Growing part   Nongrowing part

Decomposition string of complex cylinder 1

Decomposition protein spiral <<<	Decomposition protein spiral <<<
Synthetic protein spiral >>>	Decomposition protein spiral >>>

Nongrowing part   Growing part

Synthetic string of complex cylinder 2

Decomposition protein spiral >>>	Decomposition protein spiral >>>
<<< New decomposition protein spiral	<<< Synthetic protein spiral

Growing part   Nongrowing part

Decomposition string of complex cylinder 2

Synthetic protein spiral <<<	Synthetic protein spiral <<<
Decomposition protein spiral >>>	New synthetic protein spiral >>>

Nongrowing part   Growing part

Synthetic spirals are produced by synthetic subgroups of genes whereas decomposition spirals are produced by decomposition subgroups of genes from the same group of genes. Synthetic subgroups of genes permeate decomposition subgroups of genes in the groups and make the unity and conflict of opposites by producing opposite protein spirals.

Group of decomposition and synthetic genes repeat along the nucleic acids. The groups of specific genes are organized at the biggest groups that also repeat along nucleic acids and produce decomposition and synthetic protein spirals and strings that make the unity and conflict of opposites in complex cylinders. Complex cylinders attract each other making the unity and conflict of opposites of a cell.

Decomposition protein spirals of strings have concentrations of coenzymes in their polypeptides that provide decomposition of molecules with the help of synthetic spirals. Synthetic protein spirals of the same strings have quantitative concentrations of coenzymes in their polypeptides that synthesize molecules with the help of decomposition protein spirals. Decomposition and synthetic protein spirals of strings of complex cylinders provide decomposition and synthesis of molecules and growth and division of complex cylinders simultaneously by their unity and conflict of opposites.

In summary, complex cylinders with conflict of opposite characteristics of strings and protein spirals enable attraction of complex cylinders among themselves that enables unity of cells, opposite movements of their structures, growth, and mitosis.

Law of Passage of Quantitative into Qualitative Changes

Eukaryotic cells consist of pairs of complex cylinders with different qualities, lengths, and numbers. The polypeptides of the complex cylinders and their numbers, lengths, and quantities of their coenzymes and nonprotein hormones determine the characteristics of eukaryotic cells according to the law of the passage of quantitative into qualitative changes.

The different decomposition and synthetic protein spirals of decomposition and synthetic strings of pairs of complex cylinders in eukaryotic cells have polypeptides of different lengths; they attract different concentrations of coenzymes and nonprotein hormones that attract different concentrations of ions.

Different concentrations of ions and different lengths and numbers of complex cylinders determine the characteristics of eukaryotic cells according to the law of the passage of quantitative into qualitative changes. Characteristics of different eukaryotic cells are qualitative and quantitative biochemical contents, survival with and without air, different ways of oxidizing H+ ions, survival rates under different salinities, temperatures, and pH, utilization of molecules and sources of energy, frequency of depolarization and polarization, bioelectricity, movements, and other characteristics.

Law of the Negation of the Negation

Archaic polypeptides, the first forms of life, evolved by different forces of environments that acted as the law of the negation of the negation.

Unknown conditions in water forced archaic longer polypeptides to attract each other by ions and attract predecessors of coenzymes and nonprotein hormones that attracted ions and transformed by attracting each other in polypeptide spirals. Unknown conditions of environment made longer archaic polypeptide spirals and transformed them into archaic protein spirals.

Coils of the two protein spirals had opposite concentrations of unknown coenzymes and ions that attracted each other into helicoid spirals of archaic strings and thus made the unity and the conflict of opposites. Environmental conditions made the negation of the archaic protein spirals by archaic strings, which lived in pairs due to the exchange of protein and nucleic acid spirals. Unknown environmental conditions made the negation of the archaic strings, the archaic complex cylinders, the archaic eukaryotic cell and produced contemporary eukaryotic cells.

Phylogenetic Tree of Life

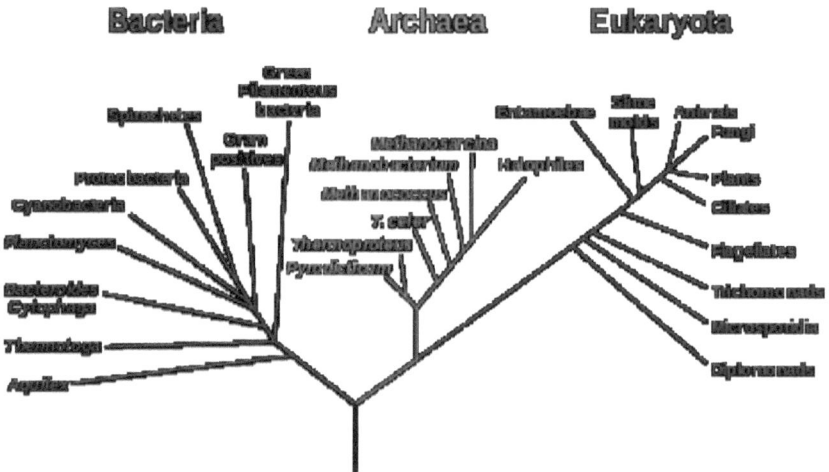

The three-domain system proposed by Woese and coworkers (1990) does not contradict the new approach to prokaryotic and eukaryotic cells though the system has a dogmatic approach. Bacteria and archaea have prokaryotic organization whereas eukaryota have eukaryotic organization according to the new approach. The system has three groups with the same origin. The three-domain system of the phylogeny model is based mainly on differences between cells in sequences of nucleotides in the cell's ribosomal RNA (r RNA).[13]

Chapter 20
Viruses and Cells

Viruses are the simplest organisms that live as parasites in prokaryotic and eukaryotic cells. They have the same origin as prokaryotic and eukaryotic cells (see appendix XXXVII).

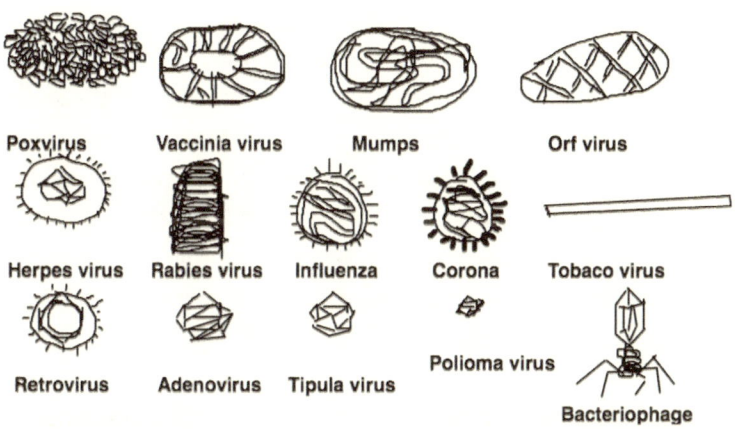

Poxvirus   Vaccinia virus   Mumps   Orf virus

Herpes virus   Rabies virus   Influenza   Corona   Tobaco virus

Retrovirus   Adenovirus   Tipula virus   Polioma virus   Bacteriophage

Shapes and sizes of viruses depend on different viral proteins and their quantities.

Viruses consist of viral proteins, nucleic acids, and sometimes nonviral molecules supplied by the host. Nonviral molecules include coenzymes, polypeptides, nonviral proteins, or different metabolites. Viral proteins consist of nonstructural proteins (nucleoproteins) and structural proteins which are produced mainly by viral nucleic acids. Nonstructural proteins consist of

•   proteins that attract viral nucleic acids,

- enzymes that help integrate viral nucleic acids with host nucleic acids as the restriction enzymes if they are DNA viruses (hypothesis), and
- enzymes that transform viral nucleic acids into opposite ones as reverse transcriptase. (RNA transform into DNA if they are ss RNA-RT viruses [+ strand or sense]—for example, retroviruses or DNA transform into RNA if they are ds DNA-RT viruses [e.g., hepadnaviruses].)

Structural proteins make different shapes of viruses.

Viruses replicate in strings of prokaryotic and eukaryotic cells because they do not have the structures necessary for biosynthesis of polypeptides and nucleic acids. They have the same origin as prokaryotic and eukaryotic cells according to the new approach. They lost the ability to replicate before the formation of prokaryotic and eukaryotic cells a long time ago most likely due to unknown environmental conditions. They survived in archaic strings as parasites because they used their structures and metabolites without any benefit to the host cells. They followed development of different forms of prokaryotic and eukaryotic cells including contemporary ones without dramatic changes losing the ability to metabolize and replicate. Some attracted parts of cell membranes and metabolites of prokaryotic and eukaryotic cells for their structures.

Viruses have icosahedron, ball, brick, rod, and pleomorphic shapes that have different nucleic acids that exist as single or multiple RNA or DNA pieces of single or double strands.

Bacteriophages entering prokaryotic cells give the impression that they inject viral nucleic acid (ds DNA, ss DNA, ds RNA, or ss RNA) and viral nucleoprotein into them. This phenomenon is due to mutual attraction of viral and prokaryotic molecules according to the new approach. Receptors of the strings of filaments of prokaryotic cells attract viruses (bacteriophages) for their molecules and vice versa. Receptors can be metabolites, coenzymes, free groups of amino acids, and polypeptides.

According to the new approach, cell membranes, cytoplasm, flagella, and pili of prokaryotic cells consist of filaments, which consist of the two strings that make the coil and the straight part on opposite sides. The ends of the protein spirals of the straight parts of the strings attract viral nucleoprotein with viral nucleic acids and pull them into strings according to the new approach. The reason is that viruses outside the cells are inert and get life only in the strings of prokaryotic cells.

The other protein parts of viruses, structural proteins, stay outside prokaryotic cells. The viral nucleoproteins have attractive affinity likely for the ends of protein spirals of the straight parts of the strings. The logic for this approach is that viral nucleoproteins and protein spirals move to the coil parts, where biosynthesis of polypeptides and nucleic acids occurs. Attractions among viral nucleoproteins and protein spirals also enable connections of viral nucleic acids with nucleic acids (ds DNA) of the strings.

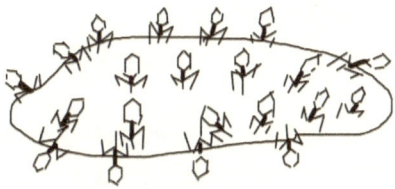

Specific receptors of the cell membranes of specific prokaryotic cells attract specific bacteriophages.

Viral nucleic acids (ds DNA, ss DNA, ds RNA, or ss RNA) attract specific strands of ds DNA of the strings by viral nucleoproteins. Viral nucleic acid can be partially or fully integrated with nucleic acids of prokaryotic cells. Fully integrated viral nucleic acids fuse with prokaryotic nucleic acids and can cause mutations of prokaryotic genes and transfer prokaryotic genes to other specific prokaryotic cells.

Viral nucleic acids and viral nucleoprotein move with ds DNA and spirals of strings to the coil parts, where biosynthesis of polypeptides occurs. The polypeptides attract viral genes that free structural viral genes for polypeptides of structural viral proteins. Free structural viral genes synthesize structural proteins of bacteriophages and viral ds RNA that transform into viral ds DNA. Structural viral proteins and viral ds DNA integrate with protein spirals of strings and their ds DNA and move along the strings to the straight parts. Structural viral proteins and viral ds DNA are expelled in the straight parts of the strings. Proteins and viral ds DNA get into the straight part of strings of the other filaments due to their attraction affinities.

Structural viral proteins, viral ds DNA, protein spirals, and string ds DNA move to the coil parts. The proteins attract genes that produce them and free nucleoprotein genes that synthesize the viral nucleoprotein and ds DNA, which move from the coil parts to the straight parts, where they are expelled.

This process of making viral proteins and ds DNA repeats, and strings accumulate viral proteins and nucleic acids. The strings break when they accumulate proteins and acids that stop metabolism of the strings due to disruption of metabolism by nonfunctional viral proteins. The affected strings release viral proteins and nucleic acids that assembled in the complete viruses due to their attractions affinities. The entrance of the viral nucleic acids into prokaryotic cells and their synthesis in prokaryotic cells are explained differently in the literature (appendix XXXIV).

The entrance of viruses in eukaryotic cells is different in comparison with prokaryotic cells due to the different structures of eukaryotic and prokaryotic cells. Different viruses are attracted by specific receptors of strings of complex cylinders of cell membranes of eukaryotic cells. Specific receptors of strings are also metabolites, coenzymes, free groups of amino acids, and polypeptides; the receptors also attract specific viruses, which are pulled in the endoplasmic reticulum and lysosomes by movements of the protein spirals. The new approach holds that this is possible only if the strings connect cell membranes with ER.

Viruses have structural and nonstructural proteins according to the literature and the new approach. Structural proteins are almost all proteins of viruses except for viral enzymes; these enzymes attract and synthesize viral nucleic acid belonging to nonstructural proteins and are known as nucleoproteins.

Enzymes of lysosomes utilize structural proteins because they do not have protection against them; the enzymes do not utilize nonstructural enzymes and viral nucleic acids because they have different enzymes with ions that repel them according to the new hypothesis. Structural proteins synthesized in affected cells are parts of the protein spirals that protect them with enzymes against destructions of lysosome enzymes; therefore, the host's structural viral proteins are not utilized.

Viral nucleoprotein and nucleic acids are attracted and can be partially or fully integrated with the ends of nongrowing parts of strings of the second or fourth cylinders in lysosomes according to their attraction affinities. Viruses with partial attractions do not fuse nucleic acids with the strings' nucleic acids. Their nucleic acids exist in strings separately despite attractions of their proteins with the strings' protein spirals.

Viruses with full attractions fuse with nucleic acids of the strings. Many viruses are partially integrated except those that cause malignant cells (adenoviruses, polyomaviruses, hepadnaviruses). Viruses with full integration can change the proto-oncogenes into oncogenes and cause malignant alterations of cells as well as mutations of different genes of eukaryotic cells.

Viral nucleoproteins and nucleic acids are partially or completely integrate with protein and nucleic acid spirals from the nongrowing to the growing parts of the strings. The different viruses with the different nucleic acids (dsDNA, dsRNA, +ss DNA, +ss RNA, -ss RNA, +ss RNA – RT, and ds DNA-RT) have similar biosynthesis of mRNA and other nucleic acids as eukaryotic cells according to the new approach.

The mRNA of viruses synthesizes the structural and nonstructural viral proteins alternatively. Polypeptides of nonstructural proteins that attract genes of nonstructural polypeptides free genes of structural polypeptides that synthesize structural polypeptides and vice versa. Synthesized structural proteins and viral nucleic acids move from the growing parts to the nongrowing parts of the strings. After the interchange of parts of protein and nucleic acid spirals, they move to the growing parts, where biosynthesis of polypeptides and nucleic acids occurs.

Polypeptides of structural proteins that attract genes of structural polypeptides free genes of nonstructural polypeptides that synthesize the nonstructural polypeptides. Nonstructural or structural polypeptides synthesize complete +ss RNA that is almost identical to mRNA. The complete +ss RNA synthesize complementary-ss RNA and ds RNA by their help. Some structural and nonstructural viral proteins of the enzymes preserve ds RNA while others transform ds RNA into viral ds DNA by string enzymes.

Viruses with single strands or viruses with ds nucleic acids always make new viral ds nucleic acids that help their multiplication. Accumulated viral proteins and nucleic acids reduce and disrupt metabolism of the strings; this disruption enables assembly of viral proteins and nucleic acids into incomplete and complete viruses. First, viral nonstructural proteins attract single-or double-stranded viral nucleic acid according to their affinities. Second, viral nonstructural proteins with single-or double-stranded nucleic acids attract structural proteins making incomplete viruses. Third, incomplete viruses attract different structures of strings or their metabolites forming complete viruses.

Some complete or incomplete viruses get out of the strings at growing or nongrowing parts with or without their destruction. The destruction or preserving strings of complex cylinders make different cytopathogenic effects inside affected eukaryotic cells.

The literature describes biosynthesis of viruses different from the way the new approach describes. (See appendix XXXV and the new approach to polypeptide and nucleic acid synthesis in chapter 11 for a better understanding of synthesis of viral polypeptides and nucleic acids.)

Viral polypeptides produce viral nucleic acids the same way as do new polypeptides of the strings. The logic in this approach is that biosynthesis of viral proteins and nucleic acids cannot differ from biosynthesis of protein spirals and nucleic acids spirals of strings because viruses use strings for biosynthesis of their proteins and nucleic acids. The same biosynthesis enables multiplication of viruses in strings.

Biosynthesis of viruses will be explained after a simple explanation of virus taxonomy because biosynthesis and assembly of viruses will be explained according to the Baltimore classification system due to its simplicity.

Viruses are classified according to the International Committee on Taxonomy of Viruses (ICTV) system and the Baltimore classification system.[14, 15] The ICTV system classifies viruses according to order (virales), family (viridae), subfamily (virinae), genus (virus), and species. There are 7 orders, 96 families, 22 subfamilies, 420 genera, and 2,618 species of viruses that have been defined by the ICTV (2012). The Baltimore classification system of viruses has seven groups:

- I: ds DNA viruses (e.g., adenoviruses, herpesviruses, poxviruses)
- II: ssDNA viruses (+ strand or "sense") DNA (e.g., parvoviruses)
- III: ds RNA viruses (e.g., reoviruses)
- IV: (+)ss RNA viruses (+ strand or sense) RNA (e.g., picornaviruses, togaviruses)
- V: (-)ss RNA viruses (-strand or antisense) RNA (e.g., orthomyxoviruses, rhabdoviruses, Ebola)
- VI: ss RNA-RT viruses (+ strand or sense) RNA with DNA intermediate in life cycle (e.g., retroviruses)
- VII: ds DNA-RT viruses DNA with RNA intermediate in life cycle (e.g., hepadnaviruses)

The Baltimore classification system of viruses is based on the way of viral mRNA synthesis. The new approach follows the Baltimore system due to easier explanations of biosynthesis of different viral proteins and nucleic acids and their assembly into viruses. See the scheme below that explains groups of viruses according to the Baltimore system.

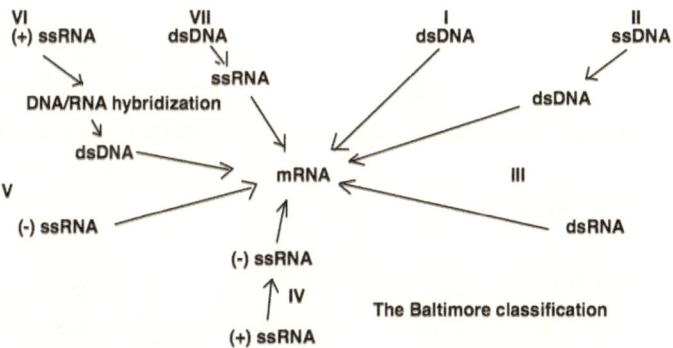

Hypothetical ways of production of the different viral nucleic acids viral proteins and their assembly into viruses are modified according to the Baltimore classification system. The hypothetical ways are these.

I. Viral ds DNA and nucleoproteins of viruses with ds DNA are fully (adenovirus) or partially (poxvirus, herpesvirus) integrated with protein spirals of the nongrowing strings of the second or fourth cylinder according to attraction affinity. The logic for this approach is that adenoviruses from this group can cause malignant transformation of affected cells by changing proto-oncogenes.

The viral ds DNAs are split also into negative ss DNA and positive ss DNA in the growing parts of the strings. Positive strands of viral ds DNA transform into mRNA. Negative strands of ss DNA synthesize positive strands of ss DNA by viral DNA replicase making new viral ds DNA. New viral ds DNA of the negative strands of viral ds DNA and new string of ds DNAs are split later into positive and negative strands of ss DNA at the growing parts of strings of complex cylinders after mitosis of eukaryotic cells.

Positive ss mRNA with viral nucleoprotein viruses synthesize viral proteins that synthesizes positive strands of RNA. Positive strands of viral nucleic acids synthesize negative strands of ss RNA making the complete viral ds RNA. Complete viral ds RNA and viral proteins transform into the complete viral ds DNA by viral enzymes during the movements of the new viral proteins and the new viral nucleic acids with protein spirals of strings. They move together from the growing to the nongrowing part. Viral proteins and nucleic acids are released at the nongrowing parts and are not utilized in lysosomes because they are integrated with parts of string protein spirals that protect them against utilization. The process repeats several times and accumulates viral proteins and nucleic acids. Viral proteins in strings reduce and stop their metabolism; that enables assembly of viral proteins and nucleic acids in complete ds DNA viruses.

II. Viral positive strand ss DNA and viral nucleoproteins of the viruses with +ss DNA are partially integrated with positive strands of ds DNA and protein spirals of the strings of the second or fourth cylinder according to attraction affinity. The logic for this approach is that parvoviruses do not cause malignant transformation of cells and are very small. The strand of +ss DNA viruses transforms directly into mRNA.

The +ss mRNA with viral nucleoprotein viruses synthesizes viral proteins that synthesize positive strands of ss RNA, which synthesize negative strands of ss RNA making viral ds RNA. Viral ds RNA and viral proteins transform into viral ds DNA during the movement of new viral proteins and nucleic acids to protein spirals that move from the growing to the nongrowing part. New viral proteins and nucleic acids are released at the nongrowing parts without assembly of complete viruses. They are not utilized in lysosomes because they are integrated with parts of the string protein spirals that protect them against utilization.

This process of making viral nucleic acids and proteins repeats several times without the assembly of viruses. Viral proteins in strings reduce and stop their metabolism, which disrupts strings at the growing parts; that enables the assembly of viral proteins and nucleic acids into complete +ss DNA viruses.

III. Viral ds RNA and nucleoproteins of the viruses with ds RNA are likely partially integrated with ds DNA and protein spirals of nongrowing strings of the second or fourth cylinder according to attraction affinity. The logic for this approach is that reoviruses do not cause malignant transformation of cells.

Viral ds RNA is not transformed into viral ds DNA because it is likely protected by viral nucleoproteins. Viral ds RNA is split into negative ss RNA and positive ss RNA in growing parts of the strings. The positive strand of viral ds RNA transforms into mRNA. The negative strand ss RNA synthesizes the positive strand ss RNA by viral RNA replicase making new viral ds RNA. New viral ds RNA of the negative strand of the viral ds DNA splits later also into positive and negative strands of ss RNA at the growing parts of the strings after mitosis of eukaryotic cells.

Positive ss mRNA with viral nucleoprotein viruses synthesize viral proteins that synthesizes positive strands of RNA. Positive strands of viral nucleic acids synthesize negative strands of ss RNA by viral RNA replicase making viral ds RNA. Viral ds RNA and viral proteins do not transform into viral ds DNA during movement through the strings to the nongrowing parts because they are protected by viral nucleoproteins from utilization and transformation into ds DNA.

This process repeats several times and accumulates viral proteins and nucleic acids. The viral proteins in strings stop their metabolism; that disrupts the strings that enable assembly of viral proteins and nucleic acids into ds RNA viruses.

IV.   Viral positive strand ss RNA and viral nucleoproteins of the viruses with +ss RNA are partially integrated with positive strands of ds DNA and protein spirals of strings of the second or fourth cylinder according to attraction affinity; picornaviruses and togaviruses do not cause malignant transformation of cells.

The strand of +ss RNA viruses transforms directly into mRNA. Positive ss mRNA with viral nucleoprotein viruses synthesize viral proteins that synthesize positive strands of ss RNA. Positive strands of ss RNA synthesize negative strands of ss RNA by viral RNA replicase making viral ds RNA. Viral ds RNA does not transform into viral ds DNA during movement because viral nucleoproteins protect them. The new viral proteins and viral ds RNA move together with protein spirals from growing to nongrowing parts of strings.

The new viral proteins and nucleic acids are released at nongrowing parts of strings without the assembly of complete viruses. They are not utilized in lysosomes because they are integrated with parts of string protein spirals that protect them against utilization. The process of making viral nucleic acids and viral proteins repeats several times without assembling viruses. Viral proteins in strings stop their metabolism and disrupt strings at the growing parts; this enables the assembly of viral proteins and nucleic acids into complete +ss RNA viruses. Viral proteins have attraction affinities only for + ss RNA and not for ds RNA.

V.   Viral negative-strand ss RNA and viral nucleoproteins of-ss RNA viruses are most likely partially integrated with negative strands of ds DNA and protein spirals of nongrowing strings of the second or fourth cylinder according to attraction affinity. The logic for this approach is that orthomyxoviruses, rhabdoviruses, and Ebola do not cause malignant transformation of cells.

The-ss RNA makes complementary +ss RNA and new ds RNA by viral RNA-dependent RNA polymerase. Viral ds RNA does not transform into viral ds DNA during movements because viral nucleoproteins preserve them. The new viral ds RNA of the negative strand of ss RNA viruses split later into positive and negative strands of ss RNA at growing parts of the strings after mitosis of eukaryotic cells. The positive strand of the ds RNA viruses transforms in mRNA as positive strands.

The positive ss mRNA with viral nucleoprotein viruses synthesizes viral proteins that synthesize positive strands of ss RNA. The positive strands of viral nucleic acids synthesize negative strands ss RNA making viral ds RNA. Viral ds RNA and proteins do not transform into viral ds DNA during movement of new viral proteins and nucleic acids with protein spirals. They move from growing to nongrowing parts of the strings. The new viral proteins and the new nucleic acids are released at the nongrowing parts without assembly.

This process repeats several times without the assembly of viruses. Viral proteins in strings stop their metabolism; this causes disruption of the strings at their growing parts and enables assembly of viral proteins and nucleic acids into complete (-ss) RNA viruses. Viral proteins have attraction affinities only for-ss RNA and not for ds RNA.

The Ebola virus belongs to the filoviridae family and to group V according to the Baltimore classification system. Ebola viruses as filaments prove the existence of strings of eukaryotic cells that synthesize them. Filaments of Ebola viruses can be born only in the strings of the complex cylinders.

Ebola virus particles attached to and budding from a chronically infected VERO E6 cell. Credit: NIAID.

The Ebola virus has negative ss RNA and structural and nonstructural proteins. Viral nucleoprotein and the protein that make complementary RNA (RNA-dependent RNA polymerase) are nonstructural proteins associated with negative ss RNA. VP24, VP40, polymerase cofactor (VP35), transcription activator (VP30), and viral envelope with glycoproteins belong to the group of structural proteins.

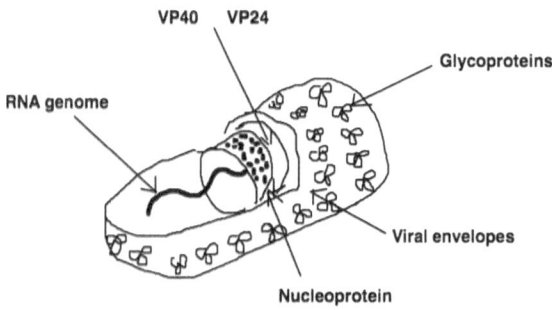

Ebola viruses belong to the negative ss RNA.

The virus is attached to the cell membrane's receptor by glycoprotein (GP). The viral envelope fuses with the cell membrane, and the nucleoprotein is released somewhere in cytoplasm according to the literature. The viral RNA-dependent RNA polymerase (RdRp, or RNA replicase) partially uncoats the nucleocapsid and transcribes the genes into positive-stranded mRNAs, which are translated into structural and nonstructural proteins according to the literature. The literature does not clearly explain the structure and biosynthesis of Ebola viruses.

VI. Viral positive strands of ss RNA, viral nucleoproteins, and reverse transcriptase are fully integrated with positive strands of ds DNA and its protein spirals of nongrowing strings of the fourth or second cylinder according to attraction affinity. The logic for this approach is that retroviruses cause malignant transformation of cells.

Reverse transcriptase (RT) transforms viral positive strands of ss RNA into viral positive strands of ss DNA according to the literature and the new approach. The viral positive strands of ss DNA transform into mRNA by string enzymes according to the new approach. Viral mRNA produces viral proteins and viral (+ss) RNA, which attracts complementary nucleotides forming complementary nucleic acids and viral ds RNA.

Viral ds RNA is fully integrated with ds RNA of strings. They are transformed into ds DNA under the influence of string enzymes. Viral protein and ds DNA move from the growing parts of the strings to the nongrowing parts, are expelled with parts of the protein spirals and nucleic acids of the strings, and move to the growing part of the strings. The ds DNA splits into positive and negative strands of ss DNA. The positive strands transform into viral mRNA that produces viral proteins and viral ds DNA.

The cycle repeats several times. The strings of the complex cylinders accumulate viral proteins and ds DNA. Viral proteins in the strings stop their metabolism and disrupt the strings at their growing parts; this enables the assembly of viral proteins and nucleic acids in the complete (+ss) RNA viruses with the reverse transcriptase enzymes. Viral proteins have attraction affinities only for-ss RNA and not for ds RNA.

VII. Viral double-stranded DNA of ds DNA-RT viruses exist as "a plasmid-like covalently closed circular DNA (cccDNA)" in the host cell nucleus according to the literature[98], but the literature does not offer any hypothesis for why viruses exist in nuclei as covalently closed circular DNA (cccDNA) and why they exist outside nuclei as relaxed circular DNA (RC–DNA). The exact mechanism for cccDNA formation is not known, although there are many papers that deal with this topic. See appendix XXXVI for more details.

The logical explanation according to the new approach is that viruses do not exist as circular ds DNA in vivo as they exist in vitro (a process taking place outside living organisms). If they exist as circular nucleic acids, they cannot make polypeptides and nucleic acids as a closed system in strings. The string structures look for the open system of viral nucleic acids for polypeptide and nucleic acid biosynthesis.

Viral ds DNA and nucleoproteins of ds DNA-RT viruses are most likely fully integrated with ds DNA and protein spirals of strings of the second or fourth cylinder according to attraction affinity. Viruses from this group can cause malignant transformation of affected cells if the viral nucleic acids are fused with proto-oncogenes.

Viral ds DNA is split on the positive strand of ss DNA and the negative strand of ss DNA in the growing parts of the strings. The positive-strand viruses transform into mRNA as positive strands of DNA of strings. Negative-strand ss DNA synthesizes positive-strand ss DNA as negative strands of strings. Positive and negative strands of viral DNA make new viral ds DNA. The new viral ds DNA of the negative strand splits later into positive and negative strands of ss DNA after mitosis of eukaryotic cells also in growing parts.

Positive mRNA with viral nucleoproteins viruses synthesize viral polypeptides and proteins that synthesize positive strands of RNA. Positive strands of viral nucleic acids synthesize negative strands of ss RNA making viral ds RNA. String proteins and nucleic acids attract viral proteins and nucleic acids. Viral ds RNA fuses with nucleic acids of strings and transforms into viral ds DNA during movement of the new viral proteins and nucleic acids to protein spirals that move from the growing part to the nongrowing part.

The new viral proteins and nucleic acids are released in the nongrowing parts. The process repeats several times and accumulates viral proteins and nucleic acids. Viral proteins in strings stop metabolism and disruptions of strings that enable assembly of viral proteins and viral nucleic acids in complete ds DNA viruses.

The new viruses break nucleic acids of strings making negative strands bigger or smaller than positive strands of viral ds DNA; this is the consequence of fusion of viral nucleic acids with string nucleic acids and attraction of positive strands by viral nucleoproteins that act also as restriction enzymes. Viral nucleoproteins as restriction enzymes cut positive and negative strands with damaging of genes including proto-oncogenes. The virus transfers damaged proto-oncogenes to new strings or new cells. The damaged complementary proto-oncogenes cause the malign transformation of the affected cells. The process of replication of viruses is explained differently in the literature (see appendix XXXV).

Cytopathogenic effect CPE in the immortal cells (for example, HELA cells) caused by viruses can be seen in the unstained and unfixed cells in Petri dishes under low-power optical microscopes.[16] The immortal cells die due to multiplication of viruses, the effect on the permeability of membrane, and the inhibition of RNA, DNA, and protein biosynthesis according to the literature. The immortal cells die due to the breaking of the strings in complex cylinders caused by the assembly of viruses that infect the cells according to the new approach. The infected strings with viruses initially cause reduction and malfunction of metabolism in them, and later, with their assembly in the complete viruses, they break the strings and complex cylinders causing different cytopathogenic effects. CPE of the infected cells manifests as

- total destruction,
- subtotal destruction,
- focal degeneration,
- swelling and clumping,
- foamy degeneration (vacuolization),
- syncytium, or
- inclusion bodies.

Manifestation of CPE can be explained according to these hypotheses.

1. Total destruction of infected cells occurs when viruses infect the strings of a majority of complex cylinders and synthesize

fast due to their simple structures. Infection spreads fast among cells due to a large amount of viruses. Therefore, a majority of the immortal cells of the monolayer in Petri dishes shrink and detach from the glass. This can be seen in enterovirus infections, which are simple viruses.

2. Subtotal destruction of infected cells occurs when viruses infect strings of a minority of complex cylinders and synthesize fast due to their simple structure. Infected cells do not produce a large amount of viruses necessary for fast spreading among cells. Therefore, a minority of the immortal cells of the monolayer in Petri dishes shrink and detach from the glass early on. This can be seen in some tagovirusis, some paramyxovirusis, and some picornavirusis.

3. Focal degeneration of infected cells occurs when viruses infect strings of a minority of complex cylinders; the viruses have affinities for specific strings. The viruses do not synthesize fast due to the complex and huge viruses that cause this phenomenon. The viruses break the strings of a minority of the cylinders and stop their metabolism causing malfunction of the infected cells. Therefore, a minority of infected immortal cells of the monolayer in Petri dishes have refractile and round shapes without shrinking due to a slow dying process. Later, the infected cells detach from the glass. Focal degeneration is due to direct cell-to-cell transfer of viruses rather than diffusion through the extracellular medium according to the literature. The cells are not totally destructed and do not produce a large amount of viruses. The infected cells due to these characteristics make focal degeneration. This type can be seen in the herpes virus and poxvirus.

4. Swelling and clumping of infected cells occurs when viruses infect strings of a majority of the cylinders. The viruses have affinities for different strings and do not synthesize fast due to the complex structure. The viruses do not break the strings of the complex cylinders at the onset of infection; they slow down their metabolism. They later cause the death of infected cells, which have huge shapes due to released metabolites that attract

water. The swelling cells clump due to the attraction of viral proteins among themselves that cause clumping of infected cells. This type can be seen in adenovirus infections.

5.   The foamy degeneration (vacuolization) of infected cells occurs when viruses infect strings of a minority of cylinders. The viruses do not synthesize fast due to the complex viruses. The viruses break the strings of a minority of complex cylinders; they stop their metabolism but do not cause the death of infected cells. They cause big vacuoles due to breaking of complex cylinders. Infected cells with vacuoles die later. This type of CPE can be observed with fixation and staining of host cells. This type of CPE can be seen in retroviruses, paramyxoviruses, and flaviviruses infections.[317]

6.   Syncytium of infected cells (big cells with more nuclei) occurs when viruses infect strings of a majority of the cylinders of affected cells. The viruses have affinities for strings of different complex cylinders. The viruses do not synthesize fast due to their complex structures. The incomplete viruses do not break the infected strings of the complex cylinders, but they slow down their metabolism. The strings with incomplete viruses of close cells attract each other and fuse with affected cells in big cells with more nuclei (syncytium). A big cell has four or more nuclei depending on how many cells are fused. A big cell with fused cytoplasm without in-the-cell membranes of the fused cells proves that the cell membrane is not a separate organelle but a part of cytoplasm. This type of CPE is detected after the cells fixation and staining, although the big cell can be seen without staining. This type of CPE can be seen in herpes viruses and paramyxovirus infection. Herpes virus also causes the other type of CPE.

7.   Inclusion bodies of infected cells occur when viruses infect strings of a minority of complex cylinders. They can be seen in cells after fixation and staining. The viruses have affinities for specific strings of the cylinders. They do not synthesize fast due to their complex structure. Inclusion bodies can be single or multiple, small or large, and round or irregular in shape.

They can be present in the cytoplasm or nucleus or both. The place of inclusion body is where the viral proteins assemble in the complete virus. The inclusion bodies can be also present without an active virus, and they present viral proteins without assembly in active viruses. This type of CPE can be seen in rabies, poliovirus, adenovirus, cytomegalovirus, and other virus infection.

Some viruses cause different malignant diseases:

- Epstein-Barr Virus (EBV)—Burkitt's lymphoma
- Hepatitis B Virus (HBV)—liver cancer
- Hepatitis C Virus (HCV)—liver cancer
- Human Herpesvirus 8 (HH8)—Kaposi's sarcoma
- Human Papillomavirus (HPV)—cervical, head, neck, anal, oral, pharyngeal, and penile cancers
- Human T-cell Lymphotropic Virus 1 (HTLV)—adult T-cell leukemia
- Merkel Cell Polyomavirus—skin cancer (Merkel Cell Carcinoma)

The above viruses cause malignant alterations of cells by different ways also according to the new approach.[17]

- They cause mutation of proto-oncogenes of DNA of the host cells.
- They damage DNA of host cells by inserting their genomes. The integration of their genomes can disrupt important regulatory genes of the host cells.
- They carry altered versions of genes they have picked up from previous host cells. These altered genes cause dysregulation and can lead to cancerous growth of infected cells.

The above viruses with their own proteins cause the opposite differentiation of the affected strings of cylinders and further consequences that cause malignant alteration of affected cells as explained in the next chapter.

CHAPTER 21
CANCER CELLS AND UNIVERSAL LAWS OF NATURE

Malignant diseases are major killers despite continuous progress and huge efforts by many biomedical scientists to find effective treatments. Simple and logical explanations of the essential characteristics and pathogenesis of cancer cells are the most important for finding treatment and preventive measures against them.

The major problem is that contemporary explanations of essential characteristics and pathogenesis of cancer cells are very complex and lack simple and logical explanations. The most important reason for this is that different phenomena of eukaryotic cells have very complex explanations that produce confusion regarding pathogenesis, characteristics, and treatment of cancer cells. Therefore, different phenomena of eukaryotic cells are explained in simple and logical ways based on logical facts and hypotheses indirectly proven by causality of different phenomena.

Hypotheses are ideas or explanations we can test through study and experimentation. In science, hypotheses are very often products of boring experiments and their facts without any imagination. "A scientific hypothesis that survives experimental testing becomes a scientific theory" according to the dictionary.[18]

The purpose of this book is to initiate new ideas and experiments regarding prokaryotic and eukaryotic cells' phenomena including new treatments for malignant and infectious diseases based on new approaches to them. This chapter deals with simple and logical explanations of the pathogenesis and essential characteristics of cancer cells by the new approach to eukaryotic cells. It also presents new ideas for effective treatment of malignant diseases.

Cancer cells are born from normal cells under influence of agents. Pathogenesis of cancer cells from normal ones is not still clear because biomedical scientists who deal with this problem do not understand

- the connections among different cells' structures, phenomena, nucleic acids, and polypeptide biosynthesis,
- the complete process and plan of differentiation of stem cells into specific ones,
- the organization of genome and connections among specific genes and cells, and
- the universal laws of nature that determine characteristics of cells including malignant ones.

To understand the pathogenesis of cancer cells, we need to understand cells' structures and phenomena, how stem cells become specific cells, and how universal laws determine characteristics of cells. This is missing in the contemporary literature, but the new approach can explain these. Abnormal complex cylinders and genes and the consequences for cell metabolism show pathogenesis and characteristics of cancer cells from a new angle. Characteristics of cancer cells described in the literature indirectly prove the normal and the abnormal complex cylinders.

Hieronymus Bosch—evolution of evil

The literature offers many known causes that transform normal cells into cancer cells—physical, chemical, and biological agents that act in proto-oncogenes. Mutated proto-oncogenes transform into oncogenes that produce oncoproteins responsible for malignant alterations of cells according to the literature. There are also unknown causes that work together with the known ones causing cancers. Cancer cells are also products of the associated causes.

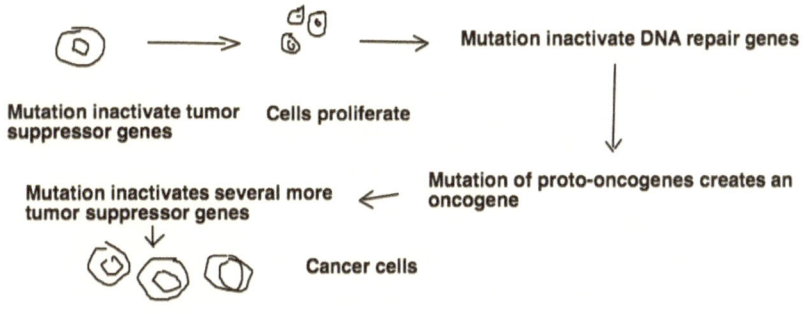

Cancer requires multiple mutations according to NIH

The new approach explains the pathogenesis of malignant cells according to the logic of the different phenomena of eukaryotic cells that are previously explained in the text. The above scheme of NIH does not explain the pathogenesis of the cancer cells according to the different phenomena of eukaryotic cells because they are not explained in the contemporary literature.[19]

The literature offers many ways that cause mutations of nucleic acids (see appendix XXXVIII).[20] They include physical causes such as ultraviolet and X-radiation as well as chemical causes such as beryllium, asbestos, aldehyde, formaldehyde, vinyl chloride, certain dioxins, aflatoxin, benzene, many substances from smoke, alkylating agents, base analogs, DNA intercalating agents, DNA cross-linkers, oxidative damage by chemicals that oxidize hydrogen and other groups of nucleotides, and many other carcinogens. Biological causes include Rous sarcoma virus, hepatitis B and C, some papillomaviruses, adenoviruses, *Helicobacter pylori, Clonorchis sinensis,* and others.

Lack of essential amino acids and high carbohydrate content in diets are major reasons for the pandemic of malignant diseases according to my opinion. Lack of essential amino acids in diets produces incomplete (broken) polypeptides that also produce mutations among proto-oncogenes because the new polypeptides synthesize nucleic acids according to the new approach.

Obese people have increased fats in different cells that complicate metabolism and increase demands for essential amino acids, which are also depleted in their diets due to different reasons (junk food with high carbohydrate content, alcohol, meat, and others). Obese people lack essential nutrients, movement, and beautiful things. Obesity is the poverty of our time.

Understanding proto-oncogenes and oncogenesis is very important for an understanding of the pathogenesis of malign cells. A proto-oncogene is a normal gene that transforms into an oncogene by mutation according to the literature and the new approach. Oncogenes produce oncoproteins that transform normal cells into malign cells. Oncoproteins cause proliferation of cells designated for apoptosis (programmed death or self-destruction) according to the literature without any logical explanations connected with the cell structures and phenomena. How oncoproteins transform normal cells into malign cells will be explained in this chapter by the new hypothesis.

Proto-oncogenes are normal genes that belong to the unknown groups of specific genes according to the new approach. Agents in specific cells change and activate proto-oncogenes according to the new approach. See the table with the proto-oncogenes and oncogenes connected with different cancer (malignant) cells in chapter 18 (Differentiation of Stem Cells).

Proto-oncogenes produce polypeptides with coenzymes. Proto-polypeptides as unknown polypeptides are actively involved in metabolism of eukaryotic cells with other polypeptides and coenzymes. Oncoproteins transform normal cells into malignant cells by unknown ways according to the literature. The authors of the literature summarized the main causes of malignancy in cells.

- DNA damage[21]
- loss of function of tumor-suppressor genes (But the literature offers no logical explanation of how suppressor genes stop

transformation of proto-oncogenes into oncogenes.)[22]

- different mutations (spontaneous mutations, mutations due to error-prone replication, bypass of naturally occurring DNA damage, errors introduced during DNA repair, and induced mutations caused by mutagens) of proto-oncogenes

These causes are accepted by the new approach to explain the pathogenesis of cancer cells.

Characteristics of Malignant Cells according to the Literature

Malignant cells have irregular sizes and shapes, and their cytoplasm is scarce. They have a large nucleus with genomic alterations, increased cell cycle and mobility, chemotaxis, changes in the cellular surface, lytic factors, invasive growth, and other characteristics.[23] The new approach to eukaryotic cells is also built according to characteristics of malign cells presented in contemporary literature, but contemporary literature does not logically explain why malignant cells have these characteristics or transformations of normal cells into malignant cells. Here are the major characteristics of malign cells with more detail according to the literature.

Cytoplasmic Membrane

- Malign cells have structural changes of proteins that do not attract corresponding ligands, new molecules that are present only in embryonic cells and increase or decrease surface receptors that increase or decrease regulatory mechanisms of the host. (The presence of new molecules present only in embryonic cells is very important because this information proves malignant cells have more qualities of different cells.)
- Malign cells with new molecules act as antigens, so they are covered with immune complexes. They have reduced acid and alkaline phosphatase and enzymes. Specific surface proteases are responsible for agglutination by plant lectins. They produce lytic factors and type IV collagenase that can destroy basal

membrane and type IV collagen. Other lytic enzymes increase their mobility and dissemination.

- Malignant cells lost contact inhibition and thus have autonomic metabolism that is increased, so they take more sugars and amino acids.
- Cells in tumors have desmosomes and junctional complexes while cells at the periphery of tumors do not have.

Cytoplasm

- They have small cytoplasm frequently with vacuoles.
- Granular endoplasmic reticulum is decreased with fragmentation and degranulation. Free ribosomes and polyribosomes are increased with an accumulation of mRNA (this makes cytoplasm basophilic).
- Agranular endoplasmic reticulum is also decreased.
- Golgi apparatus is very reduced. The cells that lost differentiation very often did not have Golgi apparatus.
- Mitochondria have different shapes and volumes with present inclusions and crystals. They have abnormal glycolysis, which is known as the Warburg phenomenon, including a cytochrome oxidase insufficiency that causes poor use of oxygen. They use glucose (five-to tenfold) more than normal cells and produce a big amount of lactic acid.
- Peroxisomes are present only in malign cells and hepatocytes. Increased numbers of peroxisomes are connected with less-differentiated cells and with increased mitosis. (Peroxisome is a small organelle present in the cytoplasm of many cells and contains the reducing enzyme catalase and usually some oxidases.)
- Glycogen is generally decreased with increased lipids except for malignant liver and kidney cells that have increased amounts.
- Secondary lysosomes appear with lipofuscin granules.
- Malignant cells have different proportions of microtubules, intermediate filaments, and microfilaments. Epithelial carcinoma has cytokeratins, and mesenchymal tumor cells have

vimentin. (Cytostatics depolymerize microtubules in malignant cells.)

- Apoptosis is present in some malignant cells; it is triggered by genetic self-destruction programs that exist in genomes of all cells.
- Malignant cells have cytoplasmic inclusions as degenerative cellular changes.

Nucleus

- The nucleus undergoes segmentation, invagination, heterochromatin reduction, increase of interchromatin and perichromatin granules, formation of inclusions, and increase of nuclear membrane pores. Cancer cells have atypical nuclei.
- Intranuclear canalicular systems form between nuclear membranes and nucleoluses. Nucleoluses are enlarged and moves to the membrane.
- Atypical mitosis increases with the defects of the mitotic spindle and atypical chromosomes.

Proliferation

- Malign cells grow continuously without local or general control. Genomic alterations are the main cause of their existence and are present in them. The above characteristics enable them to penetrate tissues causing destruction of their interstitial matrixes by lysis and their phagocytosis. They do lysis by collagenase, elastase, and proteoglycan decomposing enzymes. They have a high content of actinic filaments. Endothelial cells, fibroblasts, mastocytes, and lymphocytes of affected tissues secrete also proteolytic enzymes.
- Cells have multidirectional differentiation due to genomic alterations and therefore produce atypical substances, including collagen.

According to Carr and Underwood, malign cells cause the host to react.

- invasion of macrophage, lymphocytes and other lymphoreticular cells
- proliferation of endothelial cells and creation of new capillaries
- proliferation of fibroblasts and increased collagen production and deposition
- infiltration of neutrophils and eosinophils.

Characteristics of Malignant Cells according to the New Approach

The facts from the literature about the characteristics of malignant cells helped build the new approach to the structures and phenomena of eukaryotic cells. The most important facts are these.

- Granular endoplasmic reticulum is decreased with fragmentation and degranulation. Free ribosomes and polyribosomes are increased with an accumulation of mRNA. Malignant cells have different proportions of microtubules, intermediate filaments, and microfilaments.

These facts can be explained by the structures of the complex cylinders that build the different organelles of cytoplasm and by the disruption of the complex cylinders.

- Nucleus undergoes segmentation, invagination, heterochromatin reduction, increase of interchromatin and perichromatin granules, and formation of inclusions and increase of nuclear membrane pores. Cancer cells have atypical nuclei. Atypical mitosis increases with the defects of the mitotic spindle and atypical chromosomes.

These facts can be explained by the structures of the complex cylinders that also build the nucleus, chromosomes, and the mitotic spindle and by disrupting complex cylinders.

- New molecules of malignant cells are present only in embryonic cells. Stem, transitional, and specific cells have molecules that are products of their genomes; malignant cells have qualities of more-differentiated cells. Qualities of different cells in malignant cells can only be products of the opposite differentiation of some complex cylinders.

The new approach to structures and phenomena of eukaryotic cells can also explain how normal cells transform into malignant cells. Pathogenesis of malignant cells begins in cells exposed to mutagenic agents. Exposed cells can have high or low rates of mitosis. Those with high rates mutate faster due to their faster metabolism that produces mutated nucleic acids.

The International Agency for Research on Cancer (France) confirms that cells most commonly exposed to environmental carcinogens and cells with high rates of mitosis undergo the most common malignant alterations.

Rank	Cancer	New cases diagnosed in 2012 (1,000s)	Percent of all cancers*
1	lung	1.825	13
2	breast	1.677	11.9
3	colorectum	1.361	9.7
4	prostate	1.112	7.9
5	stomach	952	6.8
6	liver	782	5.6
7	cervix uteri	528	3.7
8	esophagus	456	3.2
9	bladder	430	3.1

10	non-Hodgkin's lymphoma	386	2.7
11	leukemia	352	2.5
12	pancreas	338	2.4
13	kidney	338	2.4
14	corpus uteri	320	2.3
15	lip, oral cavity	300	2.1
16	thyroid	298	2.1
17	brain, nervous system	256	1.8
18	ovary	239	1.7
19	melanoma	232	1.6
20	gallbladder	178	1.3
21	larynx	157	1.1
22	other pharynx	142	1.0

*Percent of all cancers excluding nonmelanoma cancers. Source: Ferlay J, Soerjomataram I, Ervik M, Dikshit R, Eser S, Mathers C, Rebelo M, Parkin DM, Forman D, Bray, F.GLOBOCAN 2012 v1.1, Cancer Incidence and Mortality Worldwide: IARC CancerBase No. 11 [internet]. Lyon, France: International Agency for Research on Cancer; 2014 (252). Available from http://globocan.iarc.fr; accessed on 1/16/2015.

Mutation of proto-oncogenes and other genes produces oncogenes. Oncogenes have the wrong nucleotides that replace existing ones or they lose nucleotides. Oncogenes cause two consequences regarding attraction of amino acids. The wrong nucleotides of the oncogenes with other close nucleotides attract wrong or no amino acids. Oncogenes in both cases produce deficient polypeptides and protein spirals.

Deficient polypeptides with changed quantities of coenzymes and nonprotein hormones and their ions cause malfunction of metabolism of affected cells. Deficient polypeptides also reduce attraction of coenzymes and nonprotein hormones. Deficient polypeptides without specific coenzymes and nonprotein hormones and their ions cannot properly attract close polypeptides and thus break strings and complex cylinders. This approach is proven by the below facts from the literature regarding the structures of the malignant cells.

- Malign cells have structural changes of proteins that do not attract corresponding ligands, the presence of new molecules that are present only in embryonic cells and increase or decrease surface receptors that increase or decrease host regulatory mechanisms.
- Malignant cells have different proportions of microtubules, intermediate filaments, and microfilaments. Epithelial carcinoma has cytokeratins, and mesenchymal tumor cells have vimentin. (Cytostatics make depolymerization of microtubules in malignant cells.)

Breaking of strings of complex cylinders causes fragmentation, degranulation, the Warburg phenomenon, different shapes of the organelles, and others. The Warburg phenomenon is due to increased metabolism as a reaction to a malfunction of metabolism of the affected cells. The result is degenerative cellular changes and apoptosis (suicide) of the affected cells. The below comments from the literature confirm the reasoning.

- Malignant cells have cytoplasmic inclusions as degenerative cellular changes.
- The granular endoplasmic reticulum is decreased with fragmentation and degranulation. Free ribosomes and polyribosomes are increased with an accumulation of mRNA; this makes cytoplasm basophilic.
- Mitochondria have different shapes and volumes with present inclusions and crystals. They have abnormal glycolysis, which is known as the Warburg phenomenon, including a cytochrome oxidase insufficiency that causes poor use of oxygen. The increased usage of glucose, five-to tenfold greater than normal cells, produces a large amount of lactic acid.
- Apoptosis, present in some malignant cells, is triggered by genetic self-destruction program that exists in the genomes of

all cells. The conclusion in the last sentence is explained the other way according to the new approach.

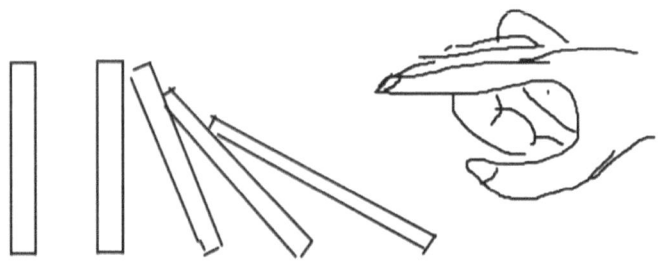

Damaged complex cylinders reduce metabolism and starve cells to death causing apoptosis in a domino effect. Self-destruction programs are unnecessary because dysfunctional metabolism kills affected cells.

The other consequence of oncopolypeptides is that their attraction for genes of specific cells changes for genes of transitional or stem cells depending on their qualities. The deficient oncopolypeptides as parts of oncoproteins cause the opposite differentiation jump in the affected complex cylinders present only in cells with mutations of proto-oncogenes and other genes. See chapter 18 where the normal differentiation jump is explained.

Oncopolypeptides initiate production of strange polypeptides that belong to transitional or stem cells. These polypeptides disrupt attraction among specific polypeptides of affected cells and produce short complex cylinders and changed complex cylinders. These short and changed cylinders cause characteristics of malignant cells—scarce cytoplasm, huge and lobulated nucleus, abnormal chromosomes, abnormal metabolism, atypical the mitotic spindle, the atypical mitosis, and others.

Normal differentiation jumps take place in complex cylinders during differentiation of stem cells into transitional and then specialized cells. The jumps take place in only one direction—from stem cells to specialized ones. The polypeptides of the strings of complex cylinders of stem cells attract groups of genes in nucleic acids of complex cylinders for transitional cells most likely due to accumulated metabolites of stem cells. These metabolites change attraction of polypeptides of stem cells for specific genes of transitional cells and cause the normal differentiation jump.

A similar phenomenon occurs in transitional cells. Polypeptides of transitional cells attract gene groups in nucleic acids of strings for specialized cells also likely due to accumulated metabolites of transitional cells. Accumulated metabolites change attraction of polypeptides of transitional cells for genes of specialized cells and cause the normal jump.

The opposite differentiation jump does not take place in specific, transitional, and stem cells under normal metabolisms, environments, polypeptides, and genes. Their protective mechanisms against opposite differentiation jumps are polypeptides in specialized cells longer than those in transitional and stem cells or longer polypeptides in transitional cells than in stem cells. The longer polypeptides with coenzymes and metabolites as ions attract only bigger specific genes because the attractive powers of the bigger genes are greater for longer polypeptides with coenzymes than for shorter ones.

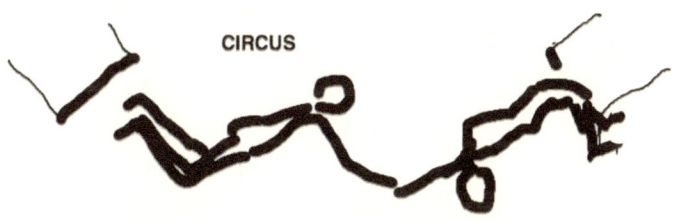

CIRCUS

The show cannot begin until you jump.

The deficient oncopolypeptides as parts of oncoproteins cause the opposite differentiation jump in the affected complex cylinders.

Opposite differentiation jumps occur only in cells with critical amounts of deficient oncopolypeptides and oncoprotein spirals of affected cylinders during their malignant alteration. The amount of the deficient oncopolypeptides below the amount of the critical one cannot causes the opposite differentiation jump in complete complex cylinders.

The facts from the literature enlighten this approach.

- Cells have multidirectional differentiation due to genomic alterations and therefore produce atypical substances including collagen.
- Malign cells grow continuously without local or general control. Genomic alterations are the main cause of their existence, and they are continuously present in them.
- Malign cells have structural changes of proteins that do not attract corresponding ligands, new molecules that are present only in embryonic cells and increase or decrease surface receptors that increase or decrease the hosts' regulatory mechanisms.

Deficient oncopolypeptides are shorter than their normal polypeptides. The shorter deficient polypeptides attract the genes of transitional polypeptides or shorter deficient polypeptides of transitional cells that attract normal specific genes of the stem polypeptides. The consequences of the opposite differentiation jumps of the affected complex cylinders of the affected cells are parts of the new complex cylinders (transitional or stem) mixed with normal ones. Parts of the new complex cylinders mixed with normal ones have shorter polypeptides and shorter cylinders compared to normal ones. The new complex cylinders in the affected complex cylinders among other ones that belong to the affected cells cause malign alterations of affected cells into malignant ones. The malignant cells have unified characteristics of atypical complex cylinders together with the consequences of broken ones and normal ones that belong to the affected cells.

The onset of malign alterations of cells with opposite differentiation jumps occurs only in small parts of the strings of the affected cylinders with oncopolypeptides and oncogenes after mitosis of affected cells, which do not show any big abnormalities microscopically at this stage. Protein spirals of affected strings have polypeptides that do not belong to specific polypeptides of the specific cells. The different polypeptides slightly change metabolism of affected complex cylinders and frequencies of depolarization or break strings of affected complex cylinders.

Carcinogens act on cell genomes and increase mutations in them. Increased mutation of nucleic acids by carcinogens in close complex cylinders takes place after many mitoses of the affected cells for a longer time. Therefore, malign alterations of the affected cells are chronic. The increased mutations cause malignant alterations of the different grades that manifest microscopically with characteristics of malignant cells. The atypical mixed and atypical pure complex cylinders produce different characteristics of malignant cells. Malignant cells with atypical complex cylinders of less mature cells have increased potential for mitosis due to increased metabolism and different coiling of the abnormal cylinders that produce earlier divisions of the affected cells.

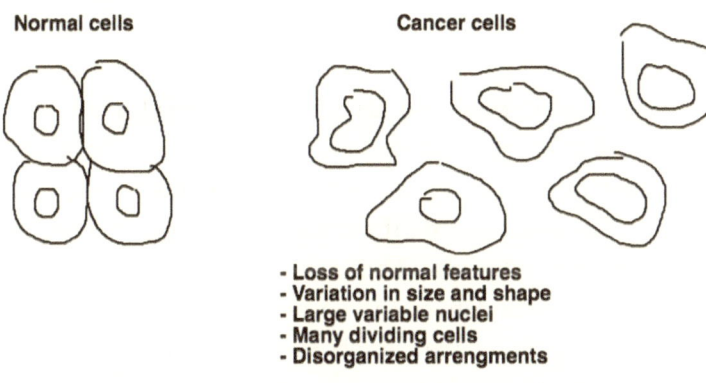

Normal cells

Cancer cells

- **Loss of normal features**
- **Variation in size and shape**
- **Large variable nuclei**
- **Many dividing cells**
- **Disorganized arrengments**

Characteristics of malign cells depend on the quantity of the atypical complex cylinders, the broken and normal ones in the affected cells. Atypical and normal complex cylinders become malignant and produce cells that resemble normal and transitional cells or stem cells depending on the transformation of the affected complex cylinders closer to the transitional complex cylinders or the stem ones. This is why malignant cells have a variety of shapes and inner structures with malignant characteristics. Malignant cells change more, and they have multidirectional differentiation due to genomic alterations because atypical complex cylinders affect biosynthesis of polypeptides of close and normal complex cylinders and cause mutations of their proto-oncogenes. Their mutated genes produce deficient polypeptides that also produce opposite differentiation jumps in the direction of less-mature cells.

Atypical complex cylinders are shorter than normal ones; they produce different metabolism with irregular mitosis of affected cells. Irregular mitosis of affected cells is due to different coiling and divisions of atypical complex cylinders in comparison with others. Irregular mitosis produces continuous changes of affected cells. The characteristics of the malignant cells from the literature are repeated in the next paragraphs for a better explanation of malignant cells' characteristics.

Malignant cells have shapes different from normal ones. They have scarce cytoplasm with huge, atypical nuclei and atypical mitosis with atypical mitotic spindles and chromosomes. The cytoplasm and the nuclei have inclusion bodies. The scarce cytoplasm has atypical mitochondria, different microtubules, and a reduced Golgi apparatus. The huge and atypical nuclei have invagination, segmentation, and bigger nuclear pores.

The mixture of normal and atypical complex cylinders increases lengths of double cylinders present in the middle of normal and abnormal cylinders, which produce huge nuclei with invagination, segmentation, and bigger nuclear pores due to different lengths and attractions among normal and atypical cylinders. The increased lengths also reduce the lengths of the four cylinders of the normal and the atypical complex cylinders producing the scarce cytoplasm. The mixture of the normal and the atypical complex cylinders produces atypical chromosomes and mitotic spindles.

Fragmentation of atypical complex cylinders manifests with fragmentation of endoplasmic reticulum and decreased granular endoplasmic reticulum with free remains of ribosomes, polyribosomes, and nucleic acid due to stopping protein and nucleic acid biosynthesis. Free nucleic acids in cytoplasm make cytoplasm basophilic. The agranular endoplasmic reticulum is also increased because affected complex cylinders are fragmented, reduced by poor metabolism, and without biosynthesis of the polypeptides and nucleic acids.

Malignant cells have asynchronous and increased metabolism with depletion of ATP, glycogen, porphyrin-Fe^{++} complexes, and fats and increased the waste products ($H+$, lactate-, $PO4---$, urea++, and others). Asynchronous and increased metabolism of malignant cells increases atypical mitosis and vulnerabilities due to lack of oxygen and metabolites and exposure to different agents (cytotoxic drugs, X-rays, and gamma rays).

Atypical and normal complex cylinders have two or more metabolisms that produce asynchronous metabolism. Atypical and normal complex cylinders fight for different metabolites because they lack them due to their asynchronous metabolism. This lack causes increased metabolism with decomposition and synthesis of metabolites. Increased metabolism of malignant cells decreases their ATP, glycogen, and fats and increases waste products and mitosis. The abnormal complex cylinders have atypical strings with less porphyrin-Fe++ complexes, which are necessary for splitting oxygen molecules in oxygen atoms for oxidation of H+ ions.

Fragmentation of complex cylinders manifest also with abnormal mitochondria of different shapes with abnormal glycolysis known as the Warburg phenomenon. Disrupted atypical complex cylinders break metabolic pathways and do not allow the making of porphyrin-Fe++ complexes for oxidation of hydrogen ions. Lack of porphyrin-Fe++ complexes manifests as oxidase insufficiency and increased concentrations of hydrogen ions in them. Malign cells produce big amounts of lactic acids due to interruption of glycolysis. Malign cells fight disrupted metabolism with increased metabolism.

A lack of glucose (energy) and oxygen molecules in malign cells can easily kill them because they need them for their increased metabolism. Cytotoxic drugs, X-rays, and gamma rays destroy normal and atypical complex cylinders with nucleic acids. They all together disturb further metabolism that kills malignant and normal cells but more malignant cells because they are more threatened. Malignant cells have irregular depolarization and bioelectricity and reduced levels of potential energy in protein spirals. They have enzyme structures different from normal ones; their enzymes have more lytic effect in molecules. Malignant cells do not have contact inhibitions that produce uncontrolled mitosis.

Normal and abnormal complex cylinders have different frequencies of depolarization that produce irregular frequencies of bioelectricity of malignant cells. This causes reduced levels of energy in protein spirals because protein spirals do not accumulate enough energy during polarization. Reduced energy in protein spirals produces very hungry decomposition enzymes that are responsible for the lytic effect of complex molecules.

The same cells of different tissues have very similar frequencies of depolarization and bioelectricities that harmonize their metabolism and growth. This harmonization causes their contact inhibition due to the influence of similar frequencies of depolarization and bioelectricities of other cells in tissues and organs. Similar frequencies of depolarization of cells in tissues of different organs produce contact inhibitions among them because they fight for the same metabolites and the other nutrients necessary for growth.

Malign cells grow independently of close cells because they have irregular depolarization and bioelectricity that are not influenced by their metabolites and bioelectricity on each other. They do not have contact inhibition by close and normal cells because they do not have synchronized metabolism with them. These above hypotheses need laboratory proofs.

In summary, the characteristics of malignant cells and the short explanations of their existences do not contradict the characteristics of malign cells described in the literature. The characteristics of malignant cells according to the literature prove the new approach to eukaryotic cells and explain the characteristics of normal and malignant cells in logical ways. Lab experiments that deal with normal and malignant cells will produce facts that will further confirm the new approach.

Malignant Tumors and Organisms in the New Approach

Malignant cells in tumors have unique metabolisms, frequencies, and energies of depolarization and bioelectricity that are not harmonized among them and the normal close cells. Malignant cells do not undergo contact inhibition due to independent and asynchronous metabolisms. They grow continuously without local or general control due to their independent and asynchronous metabolisms. Atypical and normal complex cylinders in malignant cells fight for food and due to this increase their metabolism. Malignant cells become chronically hungry and spend more glycogen and other molecules in their vicinity faster than do normal close cells. Malignant cells grow by faster supporting each other without supporting each other by their increased and asynchronous metabolisms.

Malign cells are extremely hungry due to independent and asynchronous metabolisms of complex cylinders.[24]

Malignant cells of tumors have atypical mitoses that take place also by the phenomenon of chewing gum with characteristics of malign cells. Normal and atypical complex cylinders of malign cells cause continuous growth of malign cells in tumors despite their vulnerability. The new approach can explain characteristics of malign cells while the existing literature cannot. Tumors with the above characteristics easily penetrate close tissues (lytic effects) causing tumor growth, killing close cells, damaging existing tissues, and increasing metabolites and waste products.

According to Carr and Underwood [120], malign cells cause these reactions in the host.

- invasion of macrophage, lymphocytes, and other lymphoreticular cells
- proliferation of endothelial cells and creation of new capillaries

- proliferation of fibroblasts and increased collagen production and deposition
- infiltration of neutrophils and eosinophils

Fragments of killed cells of tumors and close normal cells of tissues affect metabolism of close cells and cause inflammatory processes in affected tissues that manifest with calor, tumor, rubor, dolor, and functio laesa.

Characteristics of inflammation caused by different causes including cancer cells are summarized in the next table. Affected tissues with tumors have undergone an inflammatory process with the invasion of macrophage, neutrophils, and other lymphoreticular cells (fibroblasts) and react with proliferation of endothelial cells of their blood vessels and creation of new capillaries due to increased metabolism.

Malignant cells have abnormal proteins that cause the immune response. Cells with malign alterations due to different qualities of atypical complex cylinders have new enzymes and molecules present only in embryonic cells. Deficient polypeptides produce the immune response, so these cells are covered with immune complexes. They have reduced acid and alkaline phosphatase and changed other enzymes due to decreased energy levels of coenzymes. New enzymes as type IV collagenase (breaking of basal membranes and type IV collagen), elastase, proteoglycan-decomposing enzymes, and other proteolytic enzymes increase their mobility and dissemination.

Characteristics of inflammation caused by malignant cells	Short explanations
Increase of temperature (calor)	Affected cells fight fragments (polypeptides, parts of the protein and nucleic acids spirals and metabolites) that affect their metabolism by their increased metabolism. Cells with increased metabolism release more energy that increases the temperature of the affected tissue.
Swelling (tumor)	Increased metabolites of affected and close cells and fragments of killed cells pull in fluid from blood vessels and cause swelling of the affected tissues.
Redness (rubor)	Pulling water from blood vessels causes a better flow of blood through blood vessels that produce redness.
Pain (dolor)	Fragments of killed cells and metabolites act in nerve endings causing pain.
Changing of the function of the affected tissues (functio laesa)	Inflammatory processes in affected tissues reduce and change their functions that cause signs and symptoms of chronic diseases.

Malignant diseases are chronic diseases that kill due to disruption of the functions of the tissues of organs and attenuation of affected organisms.

Materialistic Dialectic Laws and Malignant Cells

Materialistic dialectic laws determine different characteristics of malignant cells as they do characteristics of prokaryotic and eukaryotic cells. According to the dialectic laws, malign cells have these characteristics.

I. *The law of the unity and conflict of opposites*

A malignant cell consists of normal complex cylinders of affected cells, fragmented complex cylinders, and atypical complex cylinders of different lengths. All together, they make for weak unity and conflict of extreme opposites. Atypical complex cylinders are product of accumulated mutations in normal complex cylinders of affected cells. Atypical complex cylinders are born in affected complex cylinders by opposite differentiation jumps. The normal, fragmented, and atypical complex cylinders produce specific characteristics of malign cells.

Normal and atypical cylinders of malignant cells also have two groups of protein spirals that make unity and conflict of opposites. Each group of protein spirals has decomposition and synthetic protein spirals that move opposite each other. This movement and accumulated energy in their coenzymes and other molecules enable synthesis and degradation of polypeptides and amino acids, complex and simple carbohydrates, fatty acids and fats, and nucleotides and nucleic acids.

Normal and atypical complex cylinders of affected cells have metabolites that affect their metabolisms and make for extreme conflict of opposites and weak unity. Reactions to extreme conflict of opposite metabolisms and weak unity produce their increased metabolisms and atypical metabolisms of malignant cells. If these extreme conflicts of opposite metabolisms produce stronger and weaker unity among them, malignant cells die.

Normal and atypical complex cylinders increase interchange of parts of protein and nucleic acid spirals due to increased metabolism. This increased interchange causes growth of malignant cells that is faster than that of the normal cells that produced them.

The affected complex cylinders with mutations and deficient polypeptides induce mutations in nucleic acids of close complex cylinders. As a result, malign cells change characteristics and undergo malign alterations after mitosis. In a few words, malign cells have extreme metabolic disharmonies of atypical and normal complex cylinders that increase their metabolism with extreme conflict of the opposite and weak unity.

II. *The law of the passage of quantitative changes into qualitative changes*

The quantities of polypeptides and their coenzymes in protein spirals of strings of normal and atypical complex cylinders of malign cells determine their characteristics, including these:

- different shapes
- degenerative inclusions in the cytoplasm and nucleus
- small cytoplasm and big nucleus with different shapes
- different decomposition enzymes that enable proteolysis of close cells and specific biochemical contents
- increased metabolism and proliferation
- irregular frequencies and energy of depolarization and bioelectricity

Malignant cells undergo continuous alterations that change the quantity of abnormal polypeptides and their coenzymes. This continuous change produces continuous change of the characteristics of malignant cells with disastrous effects on the cells of tissues and organs and the organisms because they weaken their metabolisms and destroy them.

III. *The law of the negation of the negation*

Malign cells negate the cells from which they are born. The previous cells negated the transitional or stem cells from which they were born. Malign cells negate the malign cells from which they were born because malign cells undergo continuous alterations that change them into new malign cells. The malign alterations are products of chronic processes of induced mutation among affected nucleic acids of the strings of affected complex cylinders of affected cells.

Treatment of Malignant Cells

Mutations take place in positive and negative strands of DNA. Positive strands undergo greater mutation than do negative ones due to their complex synthesis. Mutations in negative and positive strands take place mainly due to breaking strands by radiation, base analogs, DNA intercalating agents, DNA cross-linkers, and other ways. Mutations in positive strands take place also due to a lack of essential amino acids according to the new approach.

Due to different qualities of affected complex cylinders, malignant cells increase metabolism as a reaction to fragmentation and new qualities of complex cylinders. Malignant cells contain different metabolisms that fight each other with increasing and later decreasing metabolism due to more spending of energy and lack of metabolites. Cells with unstable metabolism are prone to make more mutations and more mitosis, and they easily die. Radiation and cytotoxic drugs kill them easily for the same reasons.

Malignant cells continuously mutate and change under the influence of radiation and cytotoxic drugs that often transform them into resistant ones. Increased mitoses and uncontrollable growth of malignant cells give them life and power. These bizarre characteristics of malign cells became gold mines for pharmaceutical companies.

2013 Global sales	Generic name	Trade name	Company	Diseases
$7.78 billion	Rituximab	Rituxan/ MabThera	Roche, Pharmstandard	non-Hodgkin's lymphoma, CLL
$6.75 billion	Bevacizumab	Avastin	Roche	colorectal, lung, ovarian, and brain cancer
$6.56 billion	Trastuzumab	Herceptin	Roche	breast, esophagus and stomach cancer
$4.69 billion	Imatinib	Gleevec	Novartis	leukemia, GI cancer
$1.09 billion	Lenalidomide	Revlimid	Celgene, Pharmstandard	multiple myeloma, mantle cell lymphoma
$2.7 billion	Pemetrexed	Alimta	Eli Lilly	lung cancer
$2.6 billion	Bortezomib	Velcade	Johnson & Johnson, Takeda, Pharmstandard	multiple myelomas
$1.87 billion	Cetuximab	Erbitux	Merck KGaA, Bristol-Myers Squibb	colon, head, and neck cancer
$1.73 billion	Leuprorelin	Lupron, Eligard	AbbVie and Takeda; Sanofi and Astellas Pharma	prostate and ovarian cancer
$1.7 billion	Abiraterone	Zytiga	Johnson & Johnson	prostate cancer

According to Wikipedia, the list of the top ten best-selling cancer drugs of 2013. The big companies are not interested in finding effective prevention and treatment of malignant diseases because they make huge amounts of money off human tragedy. They are only interested in new drugs for profits.

The above drugs cause multidirectional differentiation due more or less to mutation. Therefore, treatment with the above drugs does not have a future. There is a future only for companies that produce new drugs.

Malign cells with new qualities have characteristics of new cells in the host. Charles Darwin explained the way of surviving of different cells, including malign ones: "It is not the strongest of the species that survive, nor the most intelligent, but the one most responsive to change."

Apoptosis is present among normal and malignant cells of different tissues with damaged complex cylinders, which cannot perform normally. They reduce metabolism of affected cells, which are not strong enough to survive lack of food and influence of unnecessary metabolites with their electrical charges. Damaged complex cylinders starve cells to death causing apoptosis in a domino effect.

Apoptosis is not triggered by genetic self-destruction programs because they do not exist in genomes of cells according to the new approach. Genetic self-destruction programs are too vague to be accepted as a rationale.

Contemporary treatments of malignant cells consist of surgery, radiation, chemotherapy, hormonal treatments, and alternative methods depending on the specific malignant disease.

Future Treatment of Infectious and Malignant Diseases

Infectious and malignant diseases are a reflection of our civilization because they are products of our primitive and poor living. Different viruses and prokaryotic and eukaryotic cells cause infectious diseases of humans, animals, and plants. We fight against them by preventive measures, vaccination, and antimicrobial treatment. Malignant diseases affect humans, animals, and plants. Malignant plant cells are locked in place due to a matrix of rigid cell walls, and they do not produce metastasis.

Contemporary treatments of malignant diseases by surgery, radiation, cytotoxic, and hormone drugs slow down their development but do not cure them except in very rare cases. This is because contemporary treatments are limited due to different malignancies of tumors, widespread malignant cells through bodies (metastasis), vulnerabilities of normal cells under influence of radiation and cytotoxic drugs, and appearance of resistant malignant cells under influence of radiation and cytotoxic drugs.

Future treatments must be very efficient at killing only malignant cells. Prevention and early detection of malignant diseases will play key roles in new treatments, but preventive measures will be difficult to introduce in different societies due to poverty of the majority of people worldwide.

The enlightenment of prokaryotic, eukaryotic, and malignant cells according to the universal laws of nature has to explain their characteristics to other scientists from biological and technical fields so they can find better treatments and ways to prevent infectious and malignant diseases. Scientists who deal with bioelectricities and frequencies of depolarization of cells and use ultrasound on them will play a key role in finding new methods to detect and treat malignant diseases. This conclusion came from encouraging results of ultrasound treatment and the failure of contemporary ones.[25, 26, 27]

The literature showed electromagnetic fields and ultrasound could be used for treatment of malign tumors with different results. The works showed that "exposure to a specific time-varying Electro Magnetic Field can inhibit the growth of malignant cells."[28] They used Thomas-EMF of low intensity, frequencies modulated (25–6 Hz). Scientists who deal with ultrasound had thermal and nonthermal effects on tissues and prove malignant cells are more sensitive to thermal effects than are normal cells. The reasons for poor results of electromagnetic and ultrasound treatment of malignant tumors are poor understanding of how EMF and ultrasound affect biological processes in normal and malignant cells and poor technical devices for detection of different bioelectricity of different cells. The new approach to eukaryotic and malignant cells gives a rationale for performances of the new ultrasound and electromagnetic field experiments on cells. New experiments could build devices for successful treatment of malignant cells in vivo. The author deeply believes that EMF and ultrasound could be used together for successful treatment of malignant diseases.

Different cells and malign ones have their own frequencies of depolarization that are also frequencies of their bioelectricities because polarization and depolarization change alternatively and produce bioelectricity. Frequencies of depolarization and bioelectricity of malignant cells change due to their characteristics described in the new approach.

The new approach to eukaryotic and malignant cells can be used to come up with novel methods for detection and treatment of malignant cells not at the expense of normal cells; any object has a vibrational energy of a specific frequency. If vibrational energy is increased, an object can break apart. This approach can be explained best by the next examples.

Soldiers stop marching in rhythm when crossing a suspension bridge because the frequency of their steps might match that of the bridge and cause it to break due to resonance. Wind energy with its particular frequency can match the vibrational energy of a suspension bridge and cause a disastrous effect.

All objects in nature, including normal and malignant cells, bacteria, and viruses, have their own vibrational energy. The idea to use electromagnetic fields and ultrasound for treatment of malignant and infectious diseases is old. Dr. Royal R. Rife (1888–1971), a pioneer of this idea,[29] introduced a machine for treatment of malignant and infectious diseases. Rife's machine covered a larger frequency range—12,000 to about 20,000,000 Hz. His machine did not have practical success though he claimed to have successfully cured malignant and infectious diseases. The author of this text thinks also that Rife's machine had poor results in treatment of infectious and malignant disease because people who used it did not understand nature of normal and malignant cells. [121]

The AMA and FDA banned Rife's machine for treatment of malignant and infectious diseases for unclear reasons. His work deserves our attention despite the ban because treatment of malignant diseases by ultrasound has promising results. The below reasoning can be used for designing new experiments based on new facts regarding malignant cells and existing devices for the detection and the destruction of malignant tumors.

- Malign cells have unbalanced metabolism, frequencies, voltages, and amperages of depolarization and bioelectricities that differ from those of normal cells. The literature contains no data regarding the frequencies, voltages, and amperages of bioelectricities of malignant cells.
- Measurement of frequencies, voltages, and amperages of depolarization and bioelectricities of malign and other cells will enable the destruction of malignant cells by ultrasound treatment without hurting other cells. A new method must be developed regarding this aim. No data regarding this aim is in the literature.

Destruction of malignant cells will rely on delivering of ultrasound of specific frequencies that match those of malign cells; increasing their energy will destroy malign cells based on the concept of destructive resonance, which must be practically proven or disproven by experiments.

In summary, this book is an invitation for new ideas and methods that can treat infectious and malignant diseases. (See appendix XLII for an idea for new experimental method that requires experimentation.)

Appendices

Appendix I

Many biological departments' websites wrapped up the focus of investigations in empty words and phrases. There are many examples. One of them is this.

> Our mission is to research and teach how the collective behavior of molecules and cells forms the basis of life. We are driven by a passion for discovery and education and supported by cutting-edge research centers and state-of-the-art facilities on H ... d's C ... campus.[30]

Appendix II

The biggest problem of the dogma in biological sciences is its acceptance as absolute truth without any connections with the real nature of prokaryotic and eukaryotic cells. They defend themselves with a huge number of scientific publications that deal with dogmatic approaches to different phenomena of prokaryotic and eukaryotic cells. The worst consequence is that they stopped understanding development of the real nature of prokaryotic cells and development of the useful inventions. See the below example of a cutting-edge and state-of-the-art laboratory.[31]

> What is bioelectricity?
>
> Bioelectricity is electricity generated by living things. It is essential for many kinds of signaling in organisms ranging from bacteria to human beings. Bioelectricity

provides a mechanism for ultrarapid signaling between cells, using signaling particles readily available in the environment, i.e., charged ions. As their name suggests, proteins named 'ion channels' provide conduits for these charged ions to travel between different spaces—a common example being from the extracellular to intracellular space, or vice versa.

The inside of the cell is separated from the outside of the cell by a hydrophobic barrier formed from a lipid bilayer termed the 'plasma membrane'. Because of the hydrophobicity of their interiors, lipid bilayers are excellent insulators and prevent charged aqueous ions from freely moving between compartments. Ion channels are membrane-spanning proteins that provide a means to control the movement of the charged ions across lipid bilayers, both in terms of which ion species travel and when they travel. This is acheived without energy expenditure, as ions pass through ion channels down an electrochemical gradient, by diffusion. The mechanisms by which this is achieved are some of the most fascinating in biology. How can an ion channel distinguish between similarly charged ions, and even permit potassium ions (K+) to diffuse while disallowing the similarly charged sodium ion (Na+) which is smaller than K+? Recent advances in atomic-level structures of ion channels have revealed the secrets of their 'selectivity filters'. The ability to differentiate between K+ and Na+ is prerequisite for generation of action potentials— arguably the most important unit of signaling in the animal kingdom. In an action potential, an excitable cell, which is normally negatively charged at rest compared to the outside if the cell, first depolarizes (to a positive membrane potential) and then repolarizes (back to a negative membrane potential). The depolarization turns on processes in the cell, a common mechanism being an increase in cytosolic calcium ions (Ca2+). The repolarization returns the cell to its resting state, readying it for another signal.

The cellular depolarization is, in most excitable cells, achieved by an influx of Na+ through voltage-dependent sodium (Nav) channels. The repolarization is almost always achieved by efflux of K+ through voltage-dependent potassium (Kv) channels). Voltage dependence of ion channels is endowed by another miracle of biology—a positively charged domain that spans the lipid bilayer and senses the potential difference between the inside and the exterior of the cell. Nav and Kv voltage sensors are activated by membrane depolarization, opening the channel to permit ion diffusion in response to a change in membrane potential. Another class of channel, named "pacemaker'" or "hyperpolarization-activated" channel, is activated by hyperpolarization.

KCNE proteins

In the Abbott lab, we focus primarily on potassium channels. The pore-forming subunits of channels are typically referred to as the α subunits. These proteins contain, in Kv channels, a K+ selectivity filter and ion conduction pathway (collectively referred to as the pore) and a voltage-sensing domain (VSD). Four Kv α subunits coassemble to form a functional unit. Ion channels are however, normally macromolecular complexes made up of both α subunits and other, regulatory (β) subunits in vivo. We study how these complexes function and travel to their site of action in the cell-the plasma membrane-and what happens after they are internalized away from the plasma membrane. In particular, we study how Kv channels are regulated by a class of β or ancillary subunits encoded by the five members of the KCNE gene family. These KCNE proteins (which we also named MinK-related peptides, or MiRPs) are single-transmembrane domain proteins that co-assemble with Kv α subunits and regulate or control their trafficking (anterograde and retrograde), ion conductance and selectivity, gating kinetics (how fast or slow the channels open and close), voltage dependence (how the voltage sensor responds to membrane potential) and regulation by other proteins, lipids, and drugs.

The KCNE proteins appear to be required to generate the diversity of potassium current required for higher animals to function. We suspect that the scope of KCNE regulation of ion channels and other functions in vivo is not yet fully appreciated. Occasional hints suggest KCNE proteins may play other roles in physiology beyond regulating ion channels, or at least connecting ion channel activity to the function of other proteins or processes not currently considered to regulate channel function. Among other research directions, we are pursuing discovery of these exotic roles of KCNE proteins, utilizing techniques as diverse as transcriptomics, yeast two-hybrid, mouse genetics and electrophysiology.

The macromolecular signaling complexes we study in the Abbott lab are exemplified by those containing the KCNQ1 voltage-gated K+ (Kv) channel α subunit and KCNE family β subunits. KCNQ1 is a ubiquitous, multifunctional α subunit, genetically linked to pernicious human disorders including lethal cardiac arrhythmias, gastric cancer, and diabetes [1]. The KCNE β subunits are best known for modifying the functional properties of cardiac Kv channel α subunits including KCNQ1, and for their association with human cardiac arrhythmias including Long QT Syndrome (LQTS), Brugada Syndrome (BrS) and atrial fibrillation (AF) [2,3]. Increasing evidence supports an even broader role for KCNE subunits, encompassing most human tissues, regulation of e.g., pacemaker (HCN) and Ca2+ channels, and also the ability to dictate aspects of channel biology beyond the electrical attributes-including trafficking, α subunit composition of channels, and regulation by drugs, pH, and PIP2 [4-11]. Studies of Kcne gene knockout (Kcne–/–) mice have uncovered crucial extracardiac KCNE activities, and predicted disorders that were subsequently linked to human gene disruption [12-16]. Because the five KCNE genes exhibit non-redundant functions, and are requisite components of many native ion channels, Kcne–/– mice can be used to uncover essential native roles not just for KCNE subunits, but for KCNQ1 and other α subunits they regulate [17-19].

KCNE Genes at the Nexus of Genetic and Environmental Factors in Metabolic Disease and SCD

The majority of previous research into genes linked to sudden cardiac death (SCD) has, quite appropriately, focused on their direct roles in cardiac myocytes. However, most of the 25 genes currently associated with arrhythmia/SCD are expressed in other tissues in addition to the heart. The novel strategy adopted in the Abbott lab, highlighted by recent projects on KCNE2 and KCNE3 [17,19-22], has been to explore both the cardiac and extracardiac effects of gene disruption, and also the interplay between the two, facilitated by a holistic approach including gene-targeted mice and "omics" analyses (Figure 1).

Appendix III

Friedrich Engels and Karl Marx made revolutions in philosophy and social and natural sciences by introducing the laws of dialectic materialism and historical (evolutionary) observation of different phenomena. Engels (1820–1895) formulated the laws of the dialectical materialism as universal laws of nature in his book *Dialectics of Nature* (Wikipedia). The laws are the law of the unity and conflict of opposites, the law of the passage of quantitative into qualitative changes, and the law of the negation of the negation. They unify other known and unknown laws of nature.

Appendix IV

Charles Darwin (1809–1882) was the first biological scientist who explained evolutionary development of forms of life and the powers that changed them by the universal laws of nature. One of his famous quotes explained surviving of different species of prokaryotic and eukaryotic organisms: "It is not the strongest of the species that survive, nor the most intelligent, but the one most responsive to change."

Appendix V

Prokaryotic cells have cytoplasm covered with plasma membrane. Cytoplasm has nucleoid, plasmids, ribosomes, cytoskeletons, different proteins, polysaccharides, fats, and different granules according to contemporary microbiology. Cell bodies of prokaryotic cells are covered with gram-positive or gram-negative cell walls or do not have cell walls.

Flagella and pili get out from the cytoplasm and the plasma membrane without deeper connections with them. See the below picture of prokaryotic cells according to the contemporary literature.

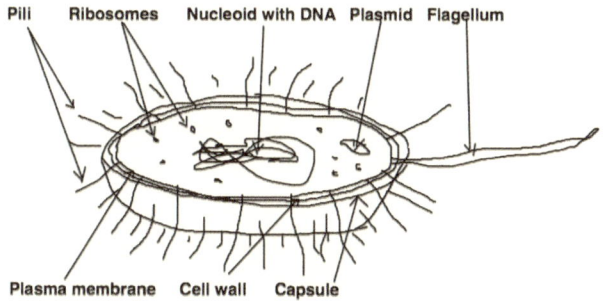

Pili   Ribosomes   Nucleoid with DNA Plasmid Flagellum

Plasma membrane   Cell wall   Capsule

The above structures float in water of the cytoplasm without any logical explanation of how they interact among themselves as a natural system. The circular chromosome of nucleoid produces mRNA in a complex and irrational way. The mRNAs move to ribosomes in an unexplained way. Metabolisms of cells takes place on unknown protein structures without logical explanations of how protein structures with enzymes interact. Many phenomena of prokaryotic cells are not explained in contemporary literature.

Understanding binary fission is impossible without understanding the inner structures of prokaryotic cells according to the new approach. These inner structures are designed to satisfy metabolism of molecules and binary fission of prokaryotic cells simultaneously. The structures and their functions are proven by metabolism of molecules and binary fission that take place simultaneously.

Binary fission in the literature means asexual reproduction of prokaryotes and some single-celled eukaryotes. Two almost identical DNAs move in opposite directions of cells after replication of genetic material without a logical explanation as to how. Cells divide into two almost equal cells also without a logical explanation. Most cells divide across the short axis. This fission is known as transverse binary fission (e.g., mostly bacilli-shaped bacteria). Some cells divide along the long axis, longitudinal binary fission (e.g., trypanosome). Some cells divide across the short axis randomly; this is known as random transverse binary fission (e.g., amoebae).[4] Some protein structures must exist to explain binary fission logically. They are not described in the literature according to the author's knowledge.

Some investigations, observations, and thoughts about binary fission with comments are in this appendix due to a better understanding of the misunderstanding regarding the structure of prokaryotic cells, binary fission, and other phenomena in contemporary literature. They are very important because they produced new ideas in reaction to their irrational approaches.

In "Electron microscopy of leptospires. I. Anatomical features of Leptospira pomona" (1965), A. E. Ritchie (National Animal Disease Laboratory, Ames, Iowa), C. Herman, and J. Ellinghausen observed *Leptospira pomona* under an electron microscope and noticed that most cells in the logarithmic phase "display a homogeneous finely granular protoplasmic cylinder." They did not notice any filaments but "some cells contain discrete or complex lamellar structures." They noticed, "The cell termini are attached to the protoplasmic cylinder by a membrane system." This fact confirmed that membrane was not a separate structure from the protoplasmic cylinder. They also noticed that "each organism appears to have two independent axial filaments originating from opposite ends of the cell." They further described that the axial filaments were composed of "a closely knit bundle of some 12 to 15 individual strands." This was a very important fact that anticipated the chewing gum phenomenon by filaments, which explains binary fission of prokaryotic cells. The chewing gum phenomenon is explained in the text. But for them, this was only suggesting a locomotor function for axial filaments, not for anything else. They wrote, "The enveloping sheath is similar to the cytoplasmic membrane; both are triple-layered and ca. 50 A thick ... The sheath preferentially forms blebs at the cell termini and midway along the protoplasmic cylinder." The blebs anticipated the interchange of protein and nucleic acid spirals among the strings of filaments. They finished without any analysis and thinking, "Septal-wall formation occurs at the mid-region of the cell, and it is concluded that simple transverse binary fission is the predominant mode of reproduction." In summary, their work anticipated the new protein structures necessary for the explanation of binary fission by the chewing gum phenomenon.[35]

In "Growth and cell division of Mycobacterium avium" (1981), authors N. Rastogi and H. L. David noticed changes in cell morphology (elongation) by optical and electron microscopy and that the rate of synthesis of RNA increased rapidly followed by an increase in protein synthesis within 3 h when the bacteria transferred from the poor to rich medium. They also noticed DNA synthesis within 7 h. The cell divisions began after a lag of about 10 h.

The ultrastructural examination of elongated cells during the adjustment period showed that they had "septa" at different stages of formation, but no evidence of fragmentation was found. They concluded that cell division in *M. avium* was by binary fission and that the notion of a lifecycle was not supported by present findings. They did not try to understand the appearance of "septa"; the magic word *septa* was enough to explain an essence of "Growth and cell division of Mycobacterium avium." Their results confirm that protein biosynthesis is connected with RNA, although they noticed DNA synthesis later.[36]

In "Population genetics of a transformable bacterium: the influence of horizontal genetic exchange on the biology of Neisseria meningitides" (1993), M. C. Maiden found that some bacterial species

> have a clonal population structure, whereas others are non-clonal or "panmictic." Clonal populations are a consequence of asexual reproduction by binary fission, whereas panmictic population structures result from 'horizontal' exchange of genetic material between clones.

Maiden unconsciously broke the model of the circular chromosome of bacteria by horizontal genetic exchange but found no explanation for this phenomenon despite the facts of mosaic gene structures that were "recognizable by comparisons of nucleotide sequences." He understood the importance that

> the human pathogen Neisseria meningitidis, several different genes, including the gene encoding the class 1 outer membrane protein, a major surface antigen, are mosaics. A consequence of horizontal genetic exchange

is that genetic process has implications both for vaccine design and in the interpretation of epidemiological data.

His fact supported the idea that bacteria had separate filaments with separate nucleic acids without circular chromosomes. If prokaryotic cells had circular chromosomes, horizontal exchange of genetic material between clones could not take place. The dogma is that the circular chromosome according to the literature is a closed system that does not allow getting in the new genetic material.[37]

In "Hypothesis: chromosomes separation in Escherichia coli involves autocatalytic gene expression, transertion and membrane domain formation" (1995), V. H. Norris proposed that bacteria had chromosomes (most likely circular) partitioned with pole preferences. He proposed that differential gene expression occurred during DNA replication in Escherichia coli. It meant according to him that daughter chromosomes had different patterns of gene expression. According to him, "differential gene expression arises from autocatalytic gene expression and creates a separate proteolipid domain around each developing chromosome via the coupled transcription-translation-insertion of proteins into membranes (transertion)." He also proposed "that the partitioning relationship between chromosome age and cell age arises because the poles of cells have a proteolipid composition that favors transertion from one nucleoid rather than from the other." If something is not explained by simple words and in a logical way, it usually means nothing in spite of hard work.[38]

In "Structure" (1996), authors M. R. J. Salton and K. S. Kim wrote an explanation of existing dogma: "Prokaryotes have a nucleoid (nuclear body) rather than an enveloped nucleus and lack membrane-bound cytoplasmic organelles." Their imagination of the circular chromosome lacked further understanding of binary fission.

Gentle lysis can be used to isolate the nucleoid of most bacterial cells. The DNA is then seen to be a single, continuous, "giant" circular molecule with a molecular weight of approximately 3 X 109. The unfolded nuclear DNA would be about 1 mm long (compared with an average length of 1 to 2 μm for bacterial cells). The bacterial nucleoid, then, is a structure containing a single chromosome.

They did not understand why "the mechanism of segregation of the two sister chromosomes following replication is not fully understood." They concluded without understanding that "models proposed require that the chromosome is permanently attached to the cell membrane throughout the various stages of the cell cycle." They wrote also this without understanding.

A flagellum consists of three parts: (1) the long filament, which lies external to the cell surface; (2) the hook structure at the end of filament; and (3) the basal body, to which the hook is anchored and which imparts motion to the flagellum. The basal body traverses the outer wall and membrane structures. It consists of a rod and one or two pairs of discs. The thrust that propels the bacterial cell is provided by counterclockwise rotation of the basal body, which causes the helically twisted filament to whirl. The movement of the basal body is driven by a proton motive force rather than by ATP directly ... Pili are more rigid in appearance than flagella.

They did not ask themselves how DNA moved through sex pili to circular chromosomes: "The sex pili attach male to female bacteria during conjugation." For them, pili are only "hairlike appendages on the surface of many gram-negative bacteria and proteins of pili are referred to as pilins." They did not try to understand the existence of the mesosomes.

Indeed, electron-microscopic studies have suggested that the mesosomes, as usually seen in thin sections, may arise from membrane perturbation and fixation artifacts. No general agreement exists about this theory,

however, and some evidence indicates that mesosomes
may be related to events in the cell division cycle.

They confirmed the confusion of the complex explanation of binary fission.[39]

In "Bipolar localization of the replication origin regions of chromosomes in vegetative and sporulation cells of Bacillus subtilis" (1997), C. D. Webb, A. Teleman, S. Gordon, et al. investigated chromosome segregation in B. subtilis and introduced "tandem copies of the lactose operon operator into the chromosome near the replication origin or terminus." Then they "visualized the position of the operator cassettes with green fluorescent protein fused to the Lac1 repressor." They noticed that "sporulating" bacteria underwent asymmetric cell division where the origins were localized near each pole of the cell whereas termini were restricted to the middle. They also noticed that "in growing cells, which undergo binary fission, origins were observed at various positions but preferentially toward the poles early in the cell cycle." The most important of their conclusion is that "these results indicate the existence of a mitotic-like apparatus that is responsible for moving the origin regions of newly formed chromosomes toward opposite ends of the cell."[40]

In "Cell reproduction cycle of Mycoplasma" (1999), M. Miyata reviewed the cell reproduction cycle of parasitic wall-free bacteria, mycoplasma. He wrote, "DNA replication of Mycoplasma capricolum started at a fixed site neighboring the dnaA gene and proceeds to both directions after a short arrest in one direction." His review does not have any logic because he did not try to find any logic: "The replicated chromosomes migrate to one and three-quarters of cell length before cell division to ensure delivery of the replicated DNA to daughter cells. The cell reproduction is based on binary fission but a branch is formed when DNA replication is inhibited." He did not explain the nature of binary fission.

The initiation frequency fits the slow speed of replication fork and DNA content is set constant. Mycoplasma pneumoniae has a terminal structure, designated as an attachment organelle, responsible for both host cell adhesion and gliding motility. The behavior of the organelle in a cell implies coupling of organelle formation to the cell reproduction cycle. Several proteins coded in three operons are delivered sequentially to a position neighboring the previous organelle and a nascent one is formed. One of the duplicated attachment organelles migrates to the opposite pole of the cell before cell division. It is becoming clear that mycoplasmas have specialized cell reproduction cycles adapted to the limited genome information and parasitic life.[41]

In "Alternative to binary fission in bacteria" (2005), E. R. Angert wrote, "Many species use alternative mechanisms, which include multiple offspring formation and budding, to reproduce." He explained this by the tantalizing images and the programs that were used conditionally.

In some bacterial species, these eccentric reproductive strategies are essential for propagation, whereas in others the programs are used conditionally. Although there are tantalizing images and morphological descriptions of these atypical developmental processes, none of these reproductive structures are characterized at the molecular genetic level ... Now, with newly available analytical techniques, model systems to study these alternative reproductive programs are being developed.[42]

In "FtsZ and division of prokaryotic cells and organelles" (2005), W. Margolin wrote, "Binary fission depends on FtsZ protein that is homologous to tubulin, the building block of the microtubule cytoskeleton in eukaryotes. It self-assembles into a membrane-associated 'ring' structure early in the division process." The author did not explain binary fission by FtsZ.[43]

In "Duplication and Segregation of the actin (MreB) cytoskeleton during the prokaryotic cell cycle" (2007), Purva Vats and Lawrence Rothfield reported that the actin (MreB) cytoskeleton of E. coli was also duplicated and segregated before cell division, ensuring the equipartition of the cytoskeletal element into the two daughter cells.[44]

In "Spatial complexity and control of a bacterial cycle" (2007), J. Colier and L. Shapiro wrote about a high degree of intracellular organization.

> Protein localization, notably of signal transduction proteins, chromosome partition proteins, and proteases, serves to coordinate cell division with chromosome replication and cell differentiation. The developmental fate of daughter cells is decided before completion of cytokinesis, via the early establishment of cell polarity by the distribution of activated signaling proteins, bacterial cytoskeleton, and landmark proteins.

They noticed the complexity of binary division but without any logical explanation.[45]

In "Ancient ESCRT and the evolution of binary fission" (2009), R. Y. Samson and S. D. Bell wrote that eukaryotic and prokaryotic orthologs of tubulin played key roles in DNA segregation and cell division processes. They concluded, "Recent studies have revealed that cell division can occur in the absence of this highly conserved protein." They noticed that "members of the hyperthermophilic crenarchaea, that lack tubulin-like proteins, undergo division by binary fission." They wrote further.

> Remarkably, Here we review how this process is dependent on archaeal homologs of the eukaryotic "endosomal sorting complex required for transport" (ESCRT) system—an apparatus that plays a pivotal role in a wide range of membrane manipulation processes. Thus, two distinct types of machinery to drive binary fission have evolved in prokaryotes—one dependent on tubulin-like proteins and one dependent on the ESCRT system.

They concluded that "two distinct types of machinery" caused binary fission without any logical explanation of the machinery for that.[46]

In "Cell division in a minimal bacterium in the absence of ftsZ" (2010), M. Lluch-Senar, E. Querol, and J. Piñol wrote,

> Our results demonstrate that ftsZ is non-essential for cell growth and reveal that, in the absence of the FtsZ protein, M. genitalium can manage feasible cell divisions and cytokinesis using the force generated by its mobile machinery. This is an alternative mechanism, completely independent of the FtsZ protein, to perform cell division by binary fission in a microorganism.

They explained binary fission phenomenon by m, kq, m, aq "mobile machinery" but without a clear explanation. They proposed that the mycoplasma cytoskeleton was

> a complex network of proteins involved in many aspects of the biology of these microorganisms, may have taken over the function of many genes involved in cell division, allowing their loss in the regressive evolution of the streamlined mycoplasma genomes ... Mycoplasma genomes exhibit an impressively low amount of genes involved in cell division and some species even lack the ftsZ gene, which is found widespread in the microbial world and is considered essential for cell division by binary fission.[47]

In "Bacterial chromosome organization and segregation" (2010), E. Toro and L. Shapiro wrote,

> Recent advances in cell imaging technology with subdiffraction resolution have revealed that the bacterial nucleoid is reliably oriented and highly organized within the cell. Such organization is transmitted from one generation to the next, by progressive segregation

of daughter chromosomes and anchoring of DNA to the cell envelope. Active segregation by a mitotic machinery appears to be common; however, the mode of chromosome segregation varies significantly from species to species ... Segregation is mediated by an active and dedicated mechanism of which we know very little ... We lack a mechanistic understanding of the requirements for maintaining chromosome orientation inside cells. Is one anchoring point enough to orient the entire chromosome, or do other sites need to be specifically positioned as well?

They expected to see the nature of binary fission without trying to connect different facts: "It is also still unclear why the chromosome is kept in a linear order within the cell, and not in the 'bowl of spaghetti' configuration assumed in the past." They were waiting for excitement: "We expect the field of chromosome organization and segregation to be active and exciting in the coming years." The fact was that they were not excited by their conclusions.[48]

In "Factors affecting daughter cells' arrangement during the early bacterial divisions" (2010), P. T. Su, P. W. Yen, et al. hypothesized, "The interaction between bacteria and the underneath substratum may affect the arrangement of the daughter bacteria." They tested this hypothesis by the bacterial division in hyaluronic acid (HA) gel and found,

The HA gel differs from agar by suppressing the typical side-by-side alignments to a rare population ... Examination of bacterial surface molecules that may contribute to the daughter cells' arrangement yielded an observation that, with disrupted lpp, the E. coli daughter cells increasingly formed non-typical patterns, i.e. neither sliding side-by-side in parallel nor forming elongated strings ... With oscillatory optical tweezers, we further demonstrated that the interaction force decreased in bacteria without Lpp, a result substantiating our notion that the side-by-side sliding phenomenon directly reflects the strength of in-situ interaction between bacteria and substratum ... Our results suggest strongly that the early cell patterning is affected by multiple interaction factors.[49]

In "Chromosome organization by a nucleoid-associated protein in live bacteria" (2011), W, Wang, G. W. Li, et al. concluded that "chromosome organization is facilitated by nucleoid-associated proteins (NAPs), but the mechanisms of action remain elusive." They found four NAPs—HU, Fis, IHF, and StpA—were largely scattered throughout the nucleoid. They wrote,

> In contrast, H-NS, a global transcriptional silencer, formed two compact clusters per chromosome, driven by oligomerization of DNA-bound H-NS through interactions mediated by the amino-terminal domain of the protein. H-NS sequestered the regulated operons into these clusters and juxtaposed numerous DNA segments broadly distributed throughout the chromosome ... Deleting H-NS led to substantial chromosome reorganization.

Their observations demonstrated that H-NS plays a key role in global chromosome organization in bacteria. These authors confirm that DNA cannot exist without proteins as two compact clusters per chromosome, although they do not have ideas how they should look in nature of different phenomena of prokaryotic cells.[50]

In "Positive control of cell division: FtsZ is recruited by SsgB during sporulation of Streptomyces" (2011), authors J. Willemse, J. W. Borst, et al. found that with bacteria that divide by binary fission, cell division starts with the polymerization of the tubulin homolog FtsZ at midcell to form a cell division scaffold (the Z ring), followed by recruitment of the other divisome components. According to them, the current view (2011) of bacterial cell division control starts from the principle of negative checkpoints that prevent incorrect Z-ring positioning.

What Are FtsZ and Incorrect Z-ring Positioning?

According to them, the prokaryotic cell division scaffold consisted of overlapping protofilaments of the tubulin homolog FtsZ (Bi and Lutkenhaus 1991), which in all prokaryotes studied so far was the first protein of the cell division machinery or divisome.

They wrote that several proteins were known to assist in septum site localization and stabilizing the Z ring, including FtsA and ZipA (Hale and de Boer 1997; Ray Chaudari 1999; Pichoff and Lutkenhause 2002), ZapA (Gueiros-Filho and Losick 2002), and SepF (Hamoen et al. 2006). The Z ring then mediated the recruitment of the cell division machinery or divisome to the midcell position (for review, see Goehring and Beckwith 2005; Adams and Errington 2009). According to them, recent evidence suggested that this is a two-step mechanism with a significant lag in bacillus between the formation of the Z ring associated with FtsA, ZapA, and EzrA and recruitment of the other components of cytokinetic machinery such as FtsL, FtsW, DivIVA (Gamba et al. 2009). The process of Z-ring (dis) assembly during division was actively controlled (for review, see Romberg and Levin 2003).

The above authors were hidden by the authors who provided evidence of positive control of cell division during sporulation of streptomyces via direct recruitment of FtsZ by the membrane-associated division component SsgB. Their vitro (dead) studies demonstrated that SsgB promoted the polymerization of FtsZ. The interactions were shown in vivo by time-lapse imaging and Forster resonance energy transfer and fluorescence lifetime imaging microscopy (FRET-FLIM) and were corroborated independently via two-hybrid studies.

As determined by fluorescence recovery after (FRAP), the turnover FtsZ protofilaments increased strongly at the time of Z-ring formation. The surprising positive control of Z-ring formation by SsgB implied the evolution of an entirely new way of Z-ring control, which might be explained by the absence of a midcell reference point in the long multi nucleoid hyphae. In turn, the localization of SsgB was mediated through the orthologous SsgA, and premature expression of the latter was sufficient to directly activate multiple Z-ring formations and hyper division at early stages of the streptomyces cell cycle.

Despite the many proteins mentioned here, still was not the crystal clear meaning of generations of Z-ring and its proteins as FtsZ, ZipA,

SepF, FtsA, ZapA, EzrA, FtsW, DivIVA, SsgB, and other ones and their interactions. They were assessed quantitatively in vivo using fluorescence lifetime imaging microscopy (FLIM), and sequence of localization of the proteins involved that was followed live via time-lapse imaging. They stayed as the groups of unknown proteins that confused a clear understanding of binary fission.[51]

In "Molecular and structural basis of ESCRT-III recruitment to membranes during archaeal cell division" (2011), R. Y. Samson, T. Obita, B. Hodgson, et al. found that members of the crenarchaeal kingdom such as sulfolobus divide by binary fission yet lack genes for the otherwise near ubiquitous tubulin and actin superfamilies of cytoskeletal proteins. Their work established that Sulfolobus homologs of the eukaryotic ESCRT-III and Vps4 components of the ESCRT machinery played an important role in Sulfolobus cell division. They identified a protein, CdvA, that was responsible for recruiting Sulfobolus ESCRT-III to membranes. Overexpression of the isolated ESCRT-III domain that interacted with CdvA results in the generation of nucleoid-free cells. Furthermore, CdvA and ESCRT-III synergized to deform archaeal membranes in vitro. According to them, the structure of the CdvA/ESCRT-III interface gave insight into the evolution of the more complex and modular eukaryotic ESCRT complex. In summary, the nature of binary fission was less important than CdvA/ESCRT structure because only this structure was important for its understanding.[52]

In "How did bacterial ancestors reproduce? Lessons from L-form cells and giant lipid vesicles: multiplications similarities between lipid vesicles and L-form bacteria" (2012), authors Y. Briers, P. Walde, et al. noticed something very interesting and proposed this.

> In possible scenarios on the origin of life, protocells represent the precursors of the first living cells. To study such "hypothetical" protocells, giant vesicles are being widely used as a simple model. Lipid vesicles can undergo complex morphological changes enabling self-reproduction such as growth, fission, and extra-and intravesicular budding. These properties of vesicular systems may in some way reflect the mechanism of

reproduction used by protocells. Moreover, remarkable similarities exist between the morphological changes observed in giant vesicles and bacterial L-form cells, which represent bacteria that have lost their rigid cell wall but retain the ability to reproduce. L-forms feature a dismantled cellular structure and are unable to carry out classical binary fission. We propose that the striking similarities in morphological transitions of L-forms and giant lipid vesicles may provide insights into primitive reproductive mechanisms and contribute to a better understanding of the origin and evolution of mechanisms of cell reproduction.

They did not try to understand binary fission and were not brave enough to admit they did not understand it. They hid their own misunderstanding with other authors: "Editor's suggested further reading in BioEssays Synthesizing artificial cells from giant unilamellar vesicles: State-of-the-art in the development of microfluidic technology Abstract."[53]

In "The bacterial Min system" (2013), V. W. Rowlett and W. Margolin wrote fancy explanations: "A mother cell giving rise to offspring usually needs to choose the site of cytokinesis carefully, as this will determine the size and shape of the daughter cells." How a mother cell chose the site of cytokinesis was very interesting.

So how does E. coli know where its middle is? Its cell poles are defined by the previous cell division, but, because E. coli grows by incorporating new cell wall and membrane uniformly along its length, the future cell division site at mid-cell is newly made and has no known pre-existing markers." They tried to understand this phenomenon: "One way to select the new mid-cell site would be to measure the distance from the two opposing cell poles, using a system that could recognize markers at those poles and define the spot furthest from both markers ... This would require that both polar markers act negatively on cell division at equivalent intensities. The result would be a concentration gradient, with the lowest concentration of the negative regulator at the cell midpoint, the greatest distance from both cell poles ... It turns out that E. coli and some other rod-shaped bacteria select their cell midpoint using such a negatively acting morphogen gradient, set up by the Min system, which is the focus of this Primer. As is true for many fascinating molecular mechanisms, the first inkling came from the behavior of cells in which this system was broken.[54]

In "Chromosome architecture is a key element of bacterial cellular organization" (2013), J. Ptacin and L. Shapiro revealed "unprecedented insight into the structural organization of bacterial chromosomes," but that was only their wish to understand the hidden secrets of nature. However, they noticed something puzzling.

Seminal studies in Vibrio cholerae and in the E. coli plasmid $pB_1 71$ showed that actively segregating parS loci follow a retracting cloud-like structure of the ATPase ParA, suggesting a conserved mitotic "pulling" mechanism for segregation (Fogel et al., 2006, Ringgaard et al., 2009).

A conserved mitotic "pulling" mechanism for segregation was due to the existence of separate filaments with their own DNA spirals that extended and shrank.

Further investigations revealed that ParB/parS complexes interact with ParA structures, stimulating ATP hydrolysis in ParA and release of ADP-containing subunits. In this way, multimeric ParB complexes follow along a shortening ParA structure via a burnt bridge Brownian ratchet-like mechanism (Fig. 2)(Fogel et al., 2006, Ptacin et al. 2010, Ringgaard et al., 2009, Vecchiarelli et al. 2010, Schofield et al. 2010, Leonard et al., 2005).

Further investigations confirmed that the extending parts of filaments ParB/parS complexes interacted with the shrinking parts of filaments ParA forming centrosomes that were crucial for forming spindle apparatus (a burnt bridge Brownian ratchet-like mechanism).[55]

No doubt the movement of a prokaryotic cell was in close connection with binary fission and other cell phenomena. Some authors explained the movement of the prokaryotic cell in these ways. In "How myxobacteria glide" (2002), C. Wolgemuth, E. Hoiczyk, D. Kaiser, and G. Oster wrote,

Although gliding always describes a slow surface-associated translocation in the direction of the cell's long axis, it can result from two very different propulsion mechanisms: social (S) motility and adventurous (A) motility. The force for S motility is generated by retraction of type 4 pili. The motility may be associated with the extrusion of slime, but evidence has been lacking, and how force might be generated has remained an enigma. Recently, nozzle-like structures were discovered in cyanobacteria from which slime emanated at the same rate at which the bacteria moved. This strongly implicates slime extrusion as a propulsion mechanism for gliding.

The above authors concluded that slime extrusion from such nozzles could account for most of the observed properties of A mobile gliding.[56]

In "The surprisingly diverse ways that prokaryotes move" (2008), K. F. Jarrel and M. J. McBride wrote,

Prokaryotic cells move through liquids or over moist surfaces by swimming, swarming, gliding, twitching or floating. An impressive diversity of motility mechanisms has evolved in prokaryotes. Movement can involve surface appendages, such as flagella that spin, pili that pull and Mycoplasma 'legs' that walk. Internal structures, such as the cytoskeleton and gas vesicles, are involved in some types of motility, whereas the mechanisms of some other types of movement remain mysterious. Regardless of the type of motility machinery that is employed, most motile microorganisms use complex sensory systems to control their movements in response to stimuli, which allows them to migrate to optimal environments.

They only described the fact of motility mechanisms in prokaryotes without any attempt to explain them. Complex sensory system "to control their movements in response to stimuli, which allows them to migrate to optimal environments" is even more mysterious. They did not mention the tumbling phenomenon and binary fission as movements.[57]

They researched the literature and found that a scientist Clifford Dobell (1912) had said, "The movements of the Spirochaetes are still surrounded in mystery." Another scientist Claes Weibull (1960) quoted Dobell: "It could be asked whether the situation has changed very much since those days." After yet another half of a century, many intriguing questions about spirochete motility remain unanswered. They think motility of spirochetes differs from the motility of most other bacteria in "that the entire bacterium is involved in translocation in the absence of external appendages." They explored motility and chemotaxis of *Borrelia burgdorferi* (Bb) as a model system. They found that Bb has

> periplasmic flagella (PFs) subterminally attached to each end of the protoplasmic cell cylinder, and surrounding the cell is an outer membrane. These internal helix-shaped PFs allow the spirochete to swim by generating backward-moving waves by rotation … High voltage TEM and Cryo-ET indicate that the ribbons wrap around the body axis (the longitudinal center of the cell, as if the cell is a sausage) in a right-handed sense (moving away from an observer, a right-handed helix

spirals clockwise [CW], and a left-handed helix spirals counterclockwise [CCW]).

How did they see PFs skeletal function? For them, bacterial cells and flagella are only elastic materials. They made an analogy with an AC motor without understanding skeletal structure and function.

> The stator, encoded by the motA and motB genes, is the motor force generator embedded in the protoplasmic cell membrane. The visualization of the stators in many species of bacteria is difficult, and this difficulty may in part be the result of the stator being dynamic and that its functional units freely interchange between the motors "circulating" in the protoplasmic cell membrane.

They made a hypothesis the stator-rotor without understanding anything: "This stator-rotor interaction evidently induces an unexpected conformational change of FliG, and this change is likely to be a fundamental mechanism for flagellar rotation (X. Zhao, T. Boquoi, M. Motaleb, J. Liu, unpublished)." The fact is that nature does not complicate matters as did the above authors, who were not astonished by its beauty and simplicity.

In "Physics of Intracellular Organization in Bacteria" (2015), N. S. Wingreen and K. C. Huang wrote,

> With the realization that bacteria achieve exquisite levels of the spatiotemporal organization has come to the challenge of discovering the underlying mechanisms. In this review, we describe three classes of such mechanisms, each of which has physical origins: the use of landmarks, the creation of higher-order structures that enable geometric sensing, and the emergence of length scales from systems of chemical reactions coupled to diffusion. We then examine the diversity of geometric cues that exist even in cells with relatively simple geometries and end by discussing both new

technologies that could drive further discovery and the implications of our current knowledge of the behavior, fitness, and evolution of bacteria. The organizational strategies described here are employed in a wide variety of systems and in species across all kingdoms of life; in many ways, they provide a general blueprint for organizing the building blocks of life.

From this paper, we can see that they do not understand the intracellular organization of bacteria and different phenomena that get out from their organization, although they were discussing both new technologies that could drive further discovery and the implications of our current knowledge of the behavior, fitness, and evolution of bacteria.[58]

Contemporary literature does not explain binary fission, cell structures, and movements as natural phenomena according to the logic of nature though some authors claimed to have explained it. See on YouTube.com the lecture of Lucy Shapiro[59] and Christine Jacobs-Wagner.[60]

The conclusion is that the cell wall, plasma membrane, division septum, cytoskeleton, flagella, pili, nucleoid, and other cell structures described in the literature are not connected with conjugation, transformation, transduction, protein and nucleic acid biosynthesis, metabolism of different molecules, polarization, depolarization, bioelectricity, and other phenomena. The different structures of a prokaryotic cell and its phenomena work together in causative relationships and therefore cannot be explained without each other.

Appendix VI

Flagella are parts of filaments according to the new approach. Each filament brings a pair of strings in which the protein string has the single chromosome. The existence of conjugation, transformation, transduction, and transposition phenomena look for this approach because only these structures can explain these phenomena. The strings of filament interchange parts of protein and nucleic acid spirals with close filaments. This approach is explained later in more detail. The different phenomena of prokaryotic cells look for open structures with separate ds DNA. Circular chromosomes cannot exist in live prokaryotic cells due to different phenomena that cannot take place if circular chromosomes exist.

Coagulated filaments in dead prokaryotic cells give an impression of a bowl of spaghetti. A flagellum pressed by the structures of the cell wall around it gives the impression of rings around flagella in electron images.[62] Circular structures were also products of coagulated proteins of the structures in dead prokaryotic cells before electron microscopy according to the new approach.

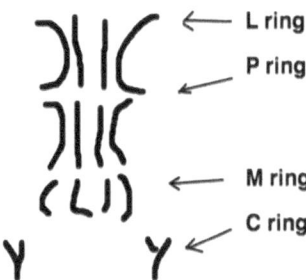

Electronic images of flagella produced tubular models of flagella.[63]

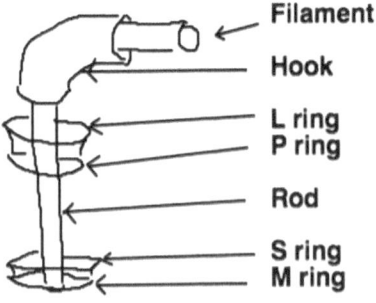

This model of flagella does not depict their real nature.[63]

Flagella as parts of filaments cannot have tubular structures as they are depicted in the drawings of contemporary models of flagella because the model cannot satisfy metabolism of molecules, conjugation, transformation, transduction, protein and nucleic acid biosynthesis, and other phenomena; it must be universal.

The tubular model was extended with the models for gram-negative and gram-positive bacteria.[64]

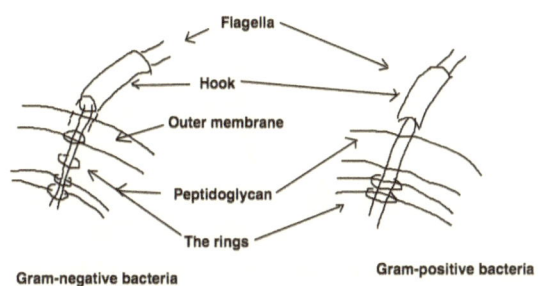

Gram-negative bacteria   Gram-positive bacteria

The extended model of flagella also did not explain the different numbers of rings at the gram-negative and the gram-positive cell walls. The flagella at the gram-negative prokaryotic cells have extra rings compared to flagella at the gram-positive prokaryotic cells despite the same structure.

"Do prokaryotes contain microtubules?" by D. Bermudes, G. Hinkle, and L. Margulis (1994) explains the nature of the protein structure inside of prokaryotic cells.[61] They present prokaryotic microtubules as two intertwined protein spirals built by lighter and darker balls. Their model provided inspiration to develop the model of a filament consisting of two strings with pairs of different and connected protein spirals. The logic of the new model is explained also in the text.

Appendix VII

Filaments according to the new approach explain the different structures of prokaryotic cells (cell membranes, ribosomes, pili, flagella, plasmids, and others) and the different phenomena of prokaryotic cells (binary fission, tumbling, protein and nucleic acid biosynthesis, and many other prokaryotic cell phenomena that carved the new approach to prokaryotic cell structure built by filaments.

Existing contemporary views of prokaryotic cells were confronted with the facts of the known structures and logic of their existence. The result was the discovery of many unknown cell structures summarized in the new model of filaments. The new structures are necessary to exist so different phenomena of prokaryotic cells can be explained logically. A good example is the logical explanation of stack structures by them.[65]

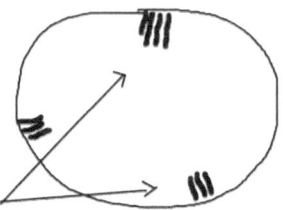

The stack structures according to electronic images of Pseudomonas deceptionensis.

> Stack structures are cuts of filaments along their lengths when they are bent in cytoplasm.[65]

Scientists noticed stack structures together with many of the circle ones (dots) in some bacterial species with transmission electron microscopy (TEM) techniques. They did not know how to explain their origin and functions because they were not brave enough to observe prokaryotic cells by an existence of filaments in cytoskeletons and their structures. Stack structures are cuts of filaments along their lengths when bent in cytoplasm while the dots are perpendicularly cuts of filaments.

Appendix VIII

Rosalind Franklin's work was crucial in the discovery of the structure of ds DNA. She made X-ray diffraction images of double strands of DNA that helped Watson and Crick build the model of ds DNA. The model of filament is called the Rosalind Franklin structure due to her contribution to the understanding of the structure of ds DNA. The ds DNA structures built the strings and their protein spirals.

Appendix IX

Specific polypeptides attract groups of coenzymes that attract metabolites. They must repeat along the polypeptide spirals of the strings of filaments (S-layers) to produce NAG and NAM molecules of peptidoglycan. The specific atom groups of NAG and NAM molecules can reconstruct coenzymes in polypeptides of close polypeptide spirals because coenzymes attract groups of atoms from them.

Appendix X

Many authors of the contemporary biology books have a similar opinion of the plasma membrane.[66]

> The plasma membrane is a thin lipid bilayer (6 to 8 nanometers) that completely surrounds the cell and separates the inside from the outside. Its selectively-permeable nature keeps ions, proteins, and other molecules within the cell, preventing them from diffusing into the extracellular environment, while other molecules may move through the membrane.

They make the similar schemes of prokaryotic cells.

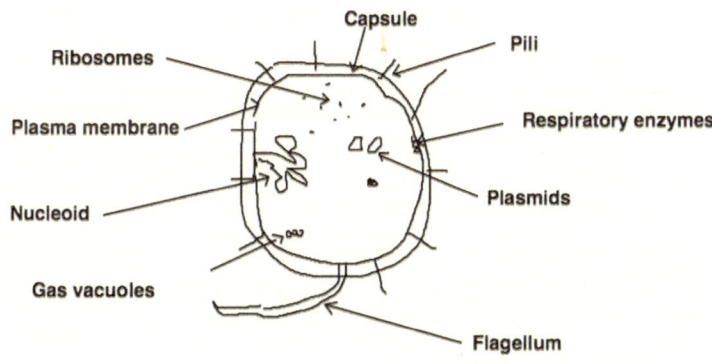

They think,

Prokaryotes are single-cell organisms that do not have a nucleus, mitochondria, or any other membrane-bound organelles. In other words, neither their DNA nor any of their other sites of metabolic activity are collected together in a discrete membrane-enclosed area. Instead, everything is openly accessible within the cell, some of which is free-floating.[66]

Some scientists who deal with the metabolism of prokaryotic and eukaryotic cells simplify their metabolism without any deeper understanding.

In general, prokaryotes lack the following membrane-bound cell compartments: mitochondria and chloroplasts. Instead, processes such as oxidative phosphorylation and photosynthesis take place across the prokaryotic plasma membrane.[67]

Appendix XI

"Enzymes are protein molecules in cells which work as catalyst. Enzymes speed up chemical reaction in the cells, but do not get used up in the process. Almost all biochemical reactions in living things need enzymes. With an enzyme, chemical reactions go much faster than they would without enzymes." 118 The International Union of Biochemistry and Molecular Biology developed a nomenclature for enzymes, the EC numbers; each enzyme is described by a sequence of four numbers preceded by EC[53, 54] The first number broadly classifies the enzyme based on its mechanism.

The top-level classification of enzymes is as follows:

- EC 1, Oxidoreductases: catalyze oxidation/reduction reactions
- EC 2, Transferases: transfer a functional group (e.g., a methyl or phosphate group)
- EC 3, Hydrolases: catalyze the hydrolysis of various bonds
- EC 4, Lyases: cleave various bonds by means other than hydrolysis and oxidation

- EC 5, Isomerases: catalyze isomerization changes within a single molecule
- EC 6, Ligases: join two molecules with covalent bonds

Appendix XII

The prokaryotic cytoskeleton is the collective name for the all structures in prokaryotic cells including filaments. FtsZ is present in almost all bacteria and regulates division. FtsZ is a tubulin homolog of eukaryotic cells. MreB and ParM regulate shape, polarity, and chromosome segregation. They are actin homologs of eukaryotic cells. CreS regulate shape; it is the homolog of an intermediate filament protein of eukaryotic cells. There are also the other proteins. MinC and MinD rapidly oscillate from pole to pole such that the time-averaged concentration of MinC is lowest at the cell middle, thereby directing FtsZ polymerization to this location (Hu and Lutkenhaus, 1999; Raskin and de Boer, 1999a; Raskin and de Boer, 1999b). Two mechanisms by which proteins are localized—diffusion capture and targeted localization—are under intense investigation in prokaryotic and eukaryotic cells. Diffusion capture requires a separate mechanism to localize the capturing factor. Targeted localization requires a mechanism to direct the targeting machinery as well as another mechanism to keep the localized factor in the correct place.[68]

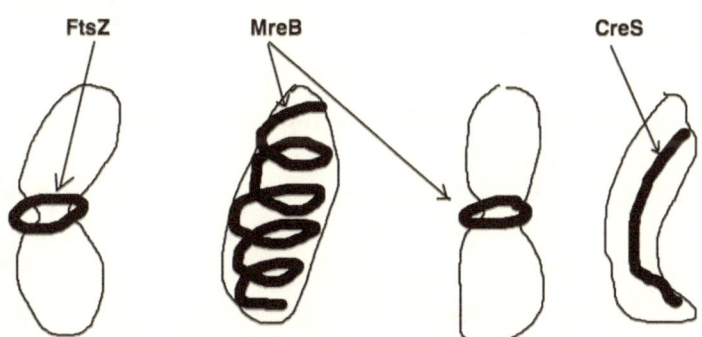

Caulobacter localization of the proteins

The contemporary model of the cytoskeleton according to the literature is based on discovered proteins.[68]

Scientists who deal with the cytoskeleton admit, "Bacteria could thus serve as models for understanding long-and short-range intercellular signaling, morphogens, cell adhesion, and tissue formation"; in other words, they do not have any idea about these phenomena.[68] They discovered many proteins in prokaryotic cells that move by the unknown mechanisms. They call them diffusion targeting and targeting localization. They expect that experiments on diffusion targeting and targeting localization will reveal these secrets without any deeper analysis of the phenomena of prokaryotic cells and an attempt to connect them in a system of causative relationships. The concept of causative relationships of phenomena of prokaryotic and eukaryotic cells according to the universal laws of nature is a powerful tool that can discover many secrets of the nature of prokaryotic and eukaryotic cells by different facts about phenomena of prokaryotic and eukaryotic cells. The most difficult task is understanding facts produced by experiments on prokaryotic cells. The new approach looks to confront the facts of causative relationships of prokaryotic and eukaryotic cells, creativity, and patience. The result is a model that is not perfect. Any model can always be improved because new facts build and destroy it.

Appendix XIII

Oxidation is the loss of electrons or an increase in oxidation state by a molecule, atom, or ion according to the literature. The reduction is the gain of electrons or a decrease in oxidation state by a molecule, atom, or ion also according to the literature. [69] The approach of the literature does not explain clear neutralization of hydrogen atoms that are the major waste of metabolism of prokaryotic and eukaryotic cells. The oxidation is neutralization of the hydrogen atoms as positive ions by oxygen atoms or by the negatively charged metabolites from Krebs cycle according to the new approach. The rationale for this approach is that low pH due to high concentrations of $H+$ ions kills prokaryotic and eukaryotic cells.

The FAD, in its fully oxidized form, or quinone form, accepts two electrons and two protons to become FADH2 (hydroquinone form) according to the literature. The semiquinone (FADH·) can be formed by either reduction of FAD or oxidation of FADH2 by accepting or donating one electron and one proton, respectively according to the literature. [70] The literature does not explain the clear function of NAD and FAD during depolarization. FADs attract hydrogen ions from NADs and release them during depolarization according to the new approach. The rationale for this approach is that some concentrations of NAD and FAD coenzymes can attract some concentrations of H+ that must be released for their neutralization.

Porphyrin-Fe++ complexes have oxido-reduction functions in prokaryotic and cells according to the literature.[71] Porphyrin-Fe++ complexes attract as the electrical positive complexes oxygen dipolar molecules or the electrical negative parts of the metabolites from the Krebs cycle according to the new approach. The rationale for this approach is that positive particles attract the negative ones and vice versa. Porphyrin – Fe++ complexes have a role in the breaking of oxygen molecules because they have abilities to attract them. The rationale is that attracted molecules can be broken only by the opposite charge. The attracted oxygen atoms or the negative metabolites from the Krebs cycle attract hydrogen atoms. The rationale for this approach is also that positive particles attract the negative ones and vice versa. The rationale is also the causative relationships among the above events.

Appendix XIV

The jumping genes are present in the prokaryotic cells[72]. If the nucleoid has closed circular ds DNA, the jumping genes cannot exist due to the circular ds DNA as closed system that does not allow any jump. The jumping genes can exist only if the ds DNA exist as pieces of the open system. The jumping genes are also a proof for the new approach to prokaryotic cells.

Appendix XV

The making of mRNA is very simple and irrational according to the literature. Messenger RNAs (mRNA) convey genetic information from DNA to the ribosomes.

The specific enzymes unzip ds DNA that codes for a specific protein. RNA polymerase binds to the promoter region of DNA and transcribes mRNA strand. Messengers RNA move to ribosomes where biosynthesis of proteins occurs according to the central dogma.

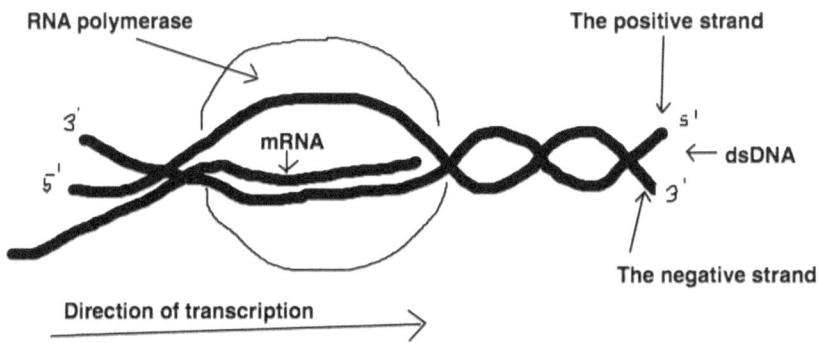

The scheme presents the model of the making of mRNA according to the literature. [73]

Appendix XVI

The nucleic acid spirals can separate along their lengths without complicated unwinding process described in the literature. The DNA helicase come from somewhere and binds to the lagging-strand template at each replication fork and moves in the 5'-> 3' direction along this strand without any explanations according to the literature.[74] It breaks hydrogen bonds and moves to the replication fork also without any explanation of how. If DNA gyrase relieves strain ahead of the replication fork without any connections with DNA helicase, there is no explanation of how they work together. In few words, this dogma does not have any logic in the real nature of the cell.

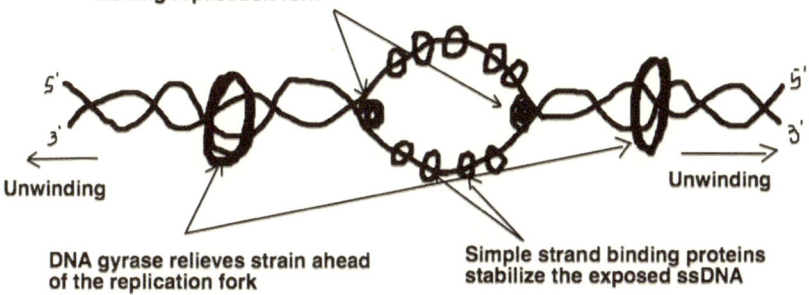

DNA helicase binds to the lagging strand template at each replication fork and moves in the 5' -> 3' direction, breaking hydrogen bonds and moving replication fork

Unwinding

Unwinding

DNA gyrase relieves strain ahead
of the replication fork

Simple strand binding proteins
stabilize the exposed ssDNA

This is an excellent example of dogma that does not have any logical explanations by the protein structures and other phenomena of the cell.

The fact is that double strands of DNA separate and open in some way to give information to the different phenomena (the polypeptide and nucleic biosynthesis, the replications of the different nucleic acids and others) of the cell. The fact is that existing dogma does not give also logical explanation how ds DNA open, because scientists did not see any connections among different cell's structures and phenomena connected with the model of the double helix of DNA. The Watson and Crick model of double-stranded DNA can satisfy and explain the different cell's phenomena only if the basic structures are found and explained by them.

Appendix XVII

The solar electric plants resemble on the porphyrin-Mg++ complex inside the strings for the collection of the Sun energy in the prokaryotic cells. The porphobilinogens of the porphyrin – Mg++ complexes have functions of the mirrors and the boiler of the solar electric plant. The porphobilinogens of the porphyrin – Mg++ complexes transfer energy to AMP via the protein spirals.

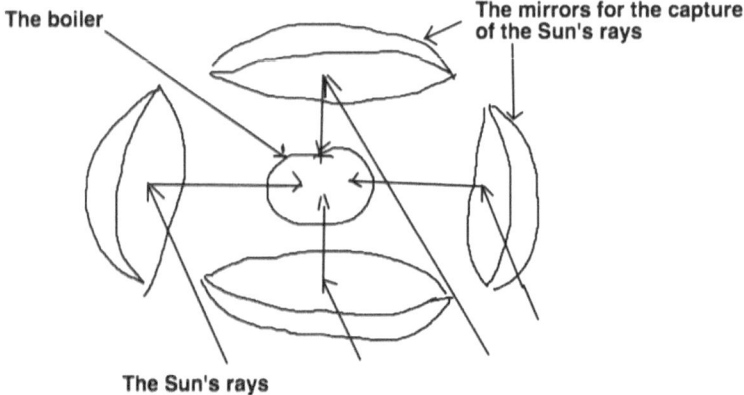

The boiler

The mirrors for the capture of the Sun's rays

The Sun's rays

Appendix XVIII

The eukaryotic cells have a very complex organization according to the contemporary literature. The different organelles of cytoplasm exist almost independently of each other without the logical explanations of their physiological connections, despite the facts that they work together.

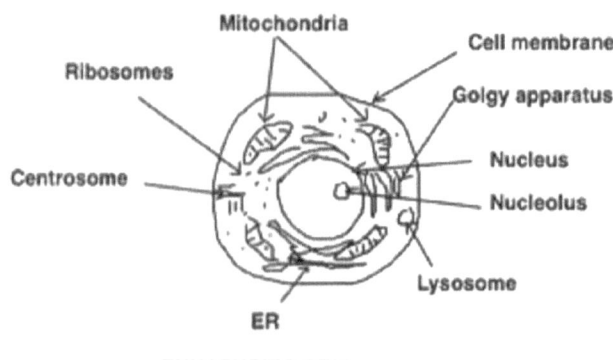

Mitochondria

Cell membrane

Ribosomes

Golgy apparatus

Centrosome

Nucleus

Nucleolus

Lysosome

ER

EUKARYOTIC CELL

The schemes of eukaryotic cells of the contemporary literature present eukaryotic cells as a bag of water in which the different organelles float as toys.

Appendix XIX

The Golgi apparatus packages "proteins into membrane-bound vesicles inside of the cell, before the vesicles are sent to their destination" according to the literature. It resides "at the intersection of the secretory, lysosomal, and endocytic pathways" also according to the literature.

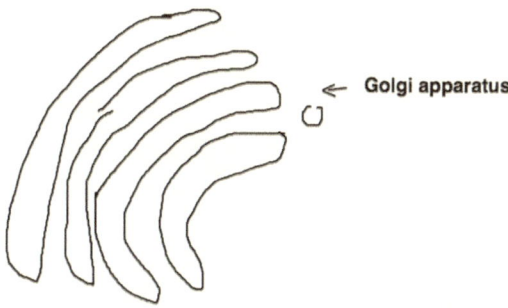

The electron images show the vesicles of the Golgi apparatus.

Appendix XX

The endoplasmic reticulum is "an interconnected network of flattened, membrane-enclosed sacs or tubes known as cisternae" according to the literature. The main functions are "the synthesis and export of proteins and membrane lipids, but its function varies between ER and cell type and cell function" also according to the literature. The endoplasmic reticulum that brings ribosomes is called rough ER, and this one that does not have them is called smooth ER.

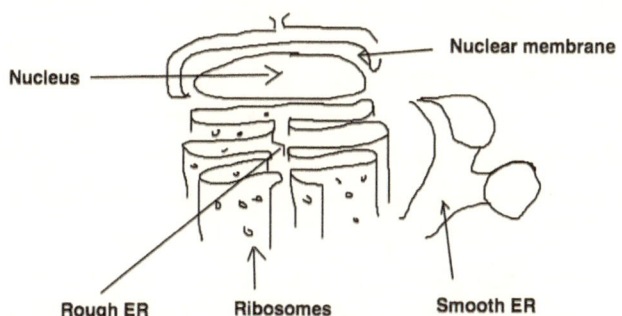

The literature's schemes of endoplasmic reticulum present ER as independent system of channels.

Appendix XXI

Eukaryotic cells are covered with a plasma membrane that is confluent and has two protein layers according to the contemporary literature. The phospholipid bilayer is between two protein layers. The plasma membrane has embedded the protein channels, the integral protein, the alpha-helix protein, glycolipid, glycoprotein, and carbohydrate. Filaments of the cytoskeleton are also connected with the inside proteins of the plasma membrane. The structures according to the existing model are presented without deeper connections how they exist in nature. See the below picture of the plasma membrane that looks as a wall made from breaks with stones and wires inside of it.

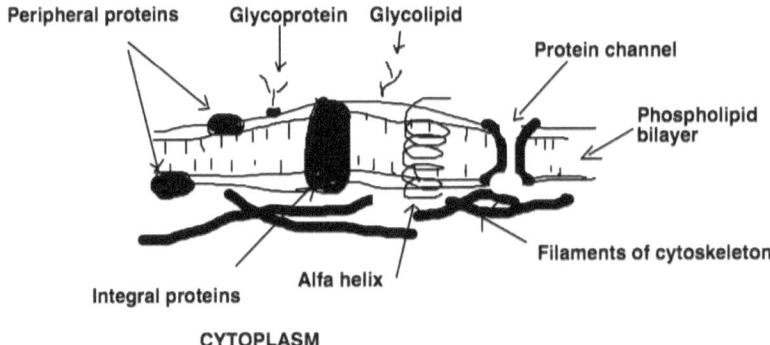

The plasma membrane looks as a wall that is built from breaks, stones, wires and other building material according to the schemes of the literature.

The plasma membrane transports different molecules according to needs of a cell by the active and the passive transport also according to the contemporary literature. The active transport enables movement of the different molecules from the outside to the inside of cell and opposite. These molecules can be huge as carbohydrates, polypeptides, fats, coenzymes and other big ones. They can also be ions as Na+, K+, Mg++, Ca++, Cl-, HPO4--, Fe++, Zn++ and other small ones. The passive transport enables mainly a movement of water and other atom ions without energy by diffusion and osmosis, due to the semipermeable membrane.

Appendix XXII

The cytoskeleton exists between the cell membrane and the nucleus. The cytoskeleton holds different organelles by a net of the tubular structures according to the literature.

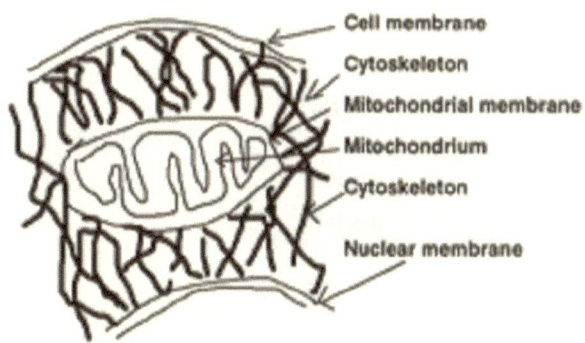

The picture does not explain origin and function of the cytoskeleton in a logical way.[166]

The cytoskeleton is "a series of intercellular proteins that help a cell with shape, support, and movement." It has microfilaments, intermediate filaments, and microtubules. No logical explanation of the origin of the different tubular structures and their functions are in the contemporary literature.

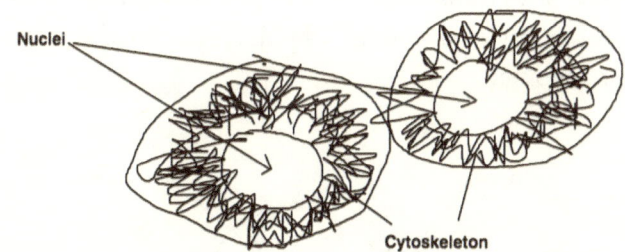

Microtubules, intermediate filaments, and microfilaments look like a chaotic mesh of wire in electronic images because these structures were coagulated during their preparation for microscopy. We cannot conclude anything except that cytoskeleton exists at eukaryotic cells.[168]

Microfilaments are the smallest of the cytoskeleton. They have G-actin protein wound in a helical shape. Intermediate filaments consist of keratin and keratin-like proteins in a cord shape. Microtubules are alpha and beta tubulin that form long hollow cylinders. They can be lengthened or shortened from the positive end.[75, 76]

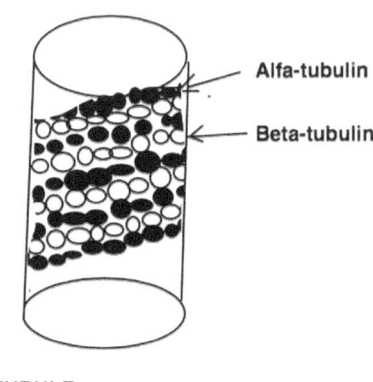

Alfa-tubulin

Beta-tubulin

MICROTUBULE

Appendix XXIII

Mitochondria "generate most of the cell's supply of ATP." They are involved in signaling, cellular differentiation, cell death, control of the cell cycle and cell growth according to the literature. The mitochondrion has its own independent genome. Further, its DNA shows substantial similarity to "bacterial genome."[77] A number of ancestral bacterial genes have also been transferred from the mitochondrial to the nuclear genome, without any explanations why the mitochondrion has its own independent genome.

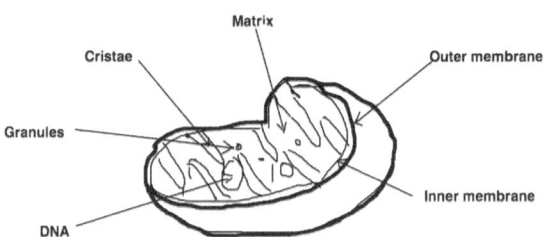

Matrix

Cristae

Granules

Outer membrane

Inner membrane

DNA

Mitochondria are also presented as independent structures according to the literature.

Appendix XXIV

Chromosomes are visible during metaphase. They exist as duplicated or unduplicated according to the literature. The unduplicated chromosomes appear as single linear strands, whereas the duplicated chromosomes contain two copies. The contemporary literature does not explain these phenomena in a logical way. Half a chromosome known as a chromatid joins to another half of chromosome or another chromatid at a protein junction known as the centromere. Chromosomes consist of single pieces of coiled ds DNA. They bring genes, regulatory elements, and noncoding DNA. Chromatin is an association of DNA with proteins and other molecules during interphase when chromosomes are not visible.

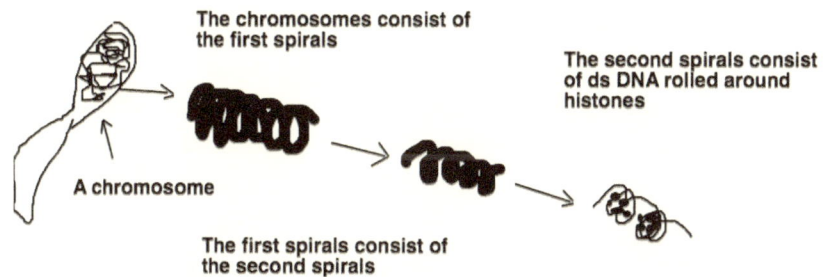

The chromosomes consist of the first spirals

The second spirals consist of ds DNA rolled around histones

A chromosome

The first spirals consist of the second spirals

THE MODEL OF A CHROMOSOME ACCORDING TO THE LITERATURE

Drawing shows connections between DNA and chromosomes according to the contemporary literature.[78]

DNA exists free and unprotected in many diagrams of the literature. DNA rolls the protein balls (histones) those coil building chromosomes. There are five different groups (families) of histones, divided into the core and the linker histones according to the literature. [79] The core histones are H2A, H2B, H3, and H4, while histones H1 and H5 are the linker histones. The two of H2A, H2B, H3, and H4 group and form an octameric nucleosome core. The big secret (according to the literature) is how histones unroll, uncoil, and prepare DNA for transcription. The big secret of functions of chromosomes will be never discovered with this approach because this picture comes out from observation of the cell with coagulated proteins. There are no explanations of logical connections among these structures because these structures do not exist in live cells, and there is no information regarding the function of all chromosomes during transcription of DNA at the same time. The chromosomes are more complex according to the new approach. There are the four protein strings organized inside of the two protein cylinders, where each protein string has DNA.

Breaking of the normal chromosome content in a normal cell is known as the chromosomal aberration. Changed numbers of chromosomes in a cell is known as aneuploidy. The gain or loss of DNA inside of chromosomes leads to genetic disorders.

Nucleic Acid

A spiral of double DNA has two complementary chains, where only a chain with genes is transcribed. The chain that is transcribed has exons or genes and introns. A gene is part of nucleic acid that synthesizes a polypeptide. Among them are parts that are spliced and not transcribed and translated into proteins. These parts are called introns. Introns are spliced (cut) and removed after those exons are joined.

There are two types of the nucleic acids DNA and RNA. DNA and RNA can be single (ss) or double stranded (ds). DNA and RNA consist of the pentose sugars, phosphates and the nucleotides.

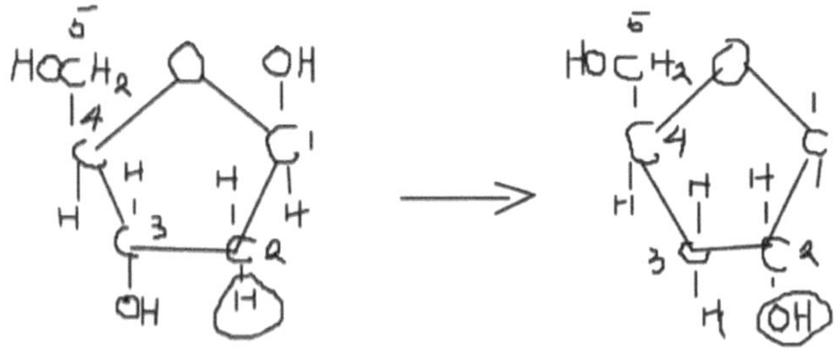

Deoxyribose

Ribose

DNA has deoxyribose whereas RNA has ribose.

The nucleotides of ds DNA or ds RNA have attractions among the complementary nucleotides (Adenine – Thymine, Guanine (Uracil) – Cytosine). RNA has Uracil instead of Guanine.

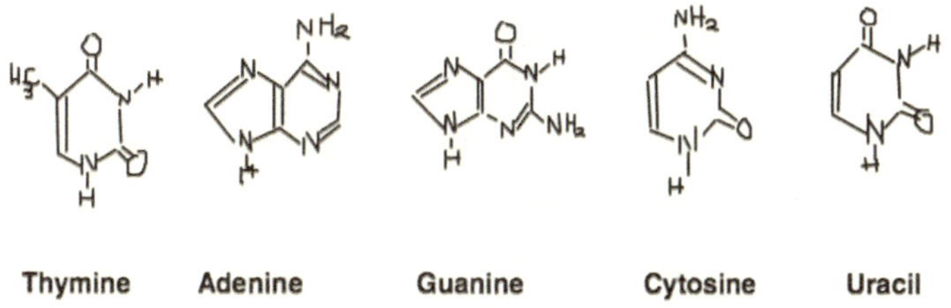

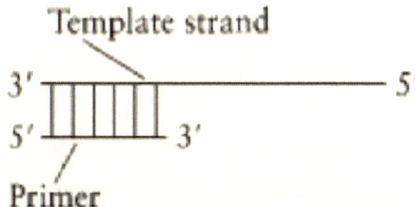

Thymine   **Adenine**   **Guanine**   **Cytosine**   **Uracil**

Nucleic acid synthesis

According to Watson and Crick, the new strands of DNA are synthesized by copying of parental strand of DNA. DNA parental strand is called template. The first copy has a complementary sequence. A copy of the copy produces the original (template) sequence again. The both parental strands of DNA are copied. The copies of the parental strands synthesize the two new duplexes.

All RNA and DNA synthesis occurs from the 5' (phosphate) end to the 3' (hydroxyl) end.

Template strand

3' ———————————— 5'

5' ———— 3'

Primer

The alpha phosphate of the incoming nucleotide attaches to the 3' hydroxyl of the ribose (deoxyribose) forming a phosphodiester bond, releasing a pyrophosphate. The chain elongation is done by pyrophosphatase that cleaves a pyrophosphate into two molecules of phosphates.

The enzymes that copy DNA are DNA polymerases. DNA polymerase cannot initiate chain synthesis without a short RNA or DNA strand, called a primer. There are a few DNA polymerases. Some of them repair DNA and some of them are present at mitosis, although it is not clear where, how, and why. The enzymes copy (transcribe) DNA to form RNA and they are called RNA polymerase. RNA polymerase finds an appropriate place in the duplex of DNA, separate and copy by "many proteins," although is not clear which and how. Only a strand is transcribed into mRNA. The double helix of DNA must unwind before transcription. The proteins called helicases do unwinding. A place of unwinding is supercoil produced by the unwinding of duplex DNA and supercoils are removed by topoisomerase. DNA replication starts with the creation of growing fork by helicases that unwind short section of the double strand of DNA.

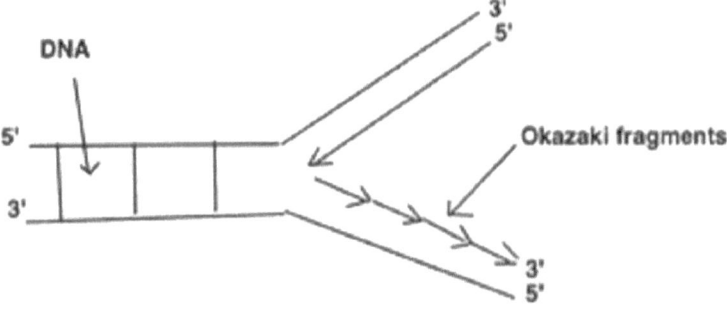

The new approach to duplication of ds DNA is made according to the contemporary model.[80]

A specialized RNA polymerase forms short RNA primers complementary to the unwound template strand (See the above picture). DNA polymerases elongate primers forming new strands. Nucleotides can be added to the growing new strands only in the 5' to 3' direction and due to this, new strands synthesize in the two opposite directions. A new strand with a direction is called the leading strand because this direction is going with an opening of the fork, without explanation of the origin of its primer. Opposite new strand is called the lagging strand. A cell produces short primers every 1000 bases or so on the second parental strand where the lagging strand is synthesized. These primers produce discontinuous fragments that are called Okazaki fragments, after their discoverer Reiji Okazaki. Finally, enzymes called ligases join these fragments. Many enzymes are involved in process of transcription of DNA and RNA. RNA processing makes functional mRNA. To the initiating (5') nucleotide of the primary transcript is added the 5' cap, which serves to protect mRNA from degradation.

Endonuclease cleaves the 3' end to yield a free 3' hydroxyl group. To the free 3' hydroxyl group is added adenylic acid residues by poly (A) polymerase. The adenylic acid residues contain 100 – 250 bases. Poly (A) polymerase is part of a complex protein that adds adenylic acid residues and it does not require a template.

Appendix XXV

The nucleus contains "most of the cell's genetic material" inside of chromosomes that appear during the division of a cell. Genetic material is organized "in long linear DNA" associated with histone proteins as chromosomes. DNAs bring genes and therefore nucleus controls the activity of the cell "by regulating gene expression." Nucleus has a double membrane with pores that regulate the nuclear transport of molecules across it. Small molecules and ions have free movements, while big proteins and RNA need carrier proteins without explanation of their origins. Nuclei also have cytoskeletons. [81]

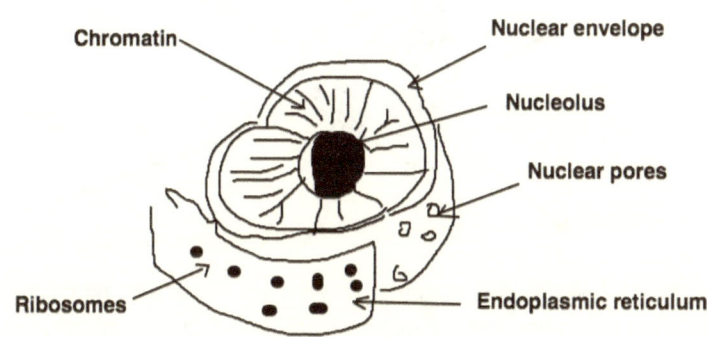

THE CONTEMPORARY MODEL OF NUCLEUS

Nucleus with nucleolus in the cytoplasm has a complex structure.[81]

Inside the nucleus exists a nucleolus as a small ball. The main roles of the nucleolus are to synthesize rRNA and assemble ribosomes.[82]

A protein called RNA polymerase I transcribe rDNA, which forms a large pre-rRNA precursor. It is cleaved into the subunits 5.8S, 18S, and 28S rRNA. The transcription, post-transcriptional processing, and assembly of rRNA occur in the nucleolus by small nucleolar RNA (snoRNA). They derived from spliced introns from mRNA encoding genes related to ribosomal function. The assembled ribosomal subunits are the largest structures passed through the nuclear pores. This approach does not explain also any deeper understanding of nucleolus.[82]

Appendix XXVI

The centrosome is the main microtubules organizing center (MTOC) of the animal cell that regulates cell division.[83] Plants and fungi do not have centrosomes and therefore use other MTOC to organize their microtubules. Centrosomes consist of two centrioles that intersect at right angles. Each centriole consists of nine triplets organized in a cartwheel structure. The pericentriolar material (PCM) as an amorphous mass of protein surrounds centrioles.

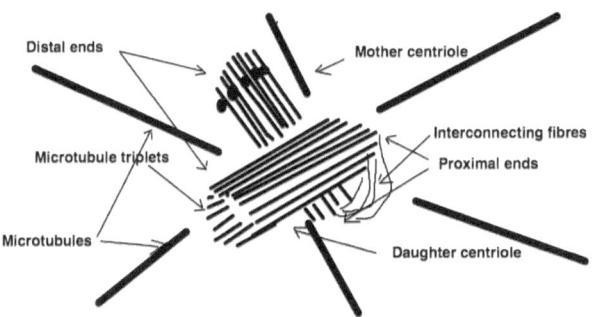

The diagram shows the contemporary model of centrosome without logical connections with the inside structures of the cytoplasm.

Centrosomes are connected with the nuclear membrane during prophase of the cell cycle. They are fused with microtubules and interact with the chromosomes building the mitotic spindle in other phases of mitosis according to the contemporary literature.

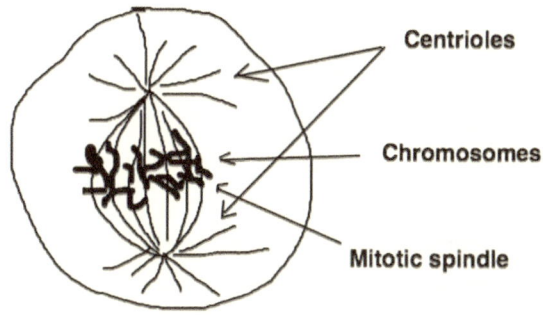

Centrioles

Chromosomes

Mitotic spindle

Chromosomes are connected with strings because strings are parts of them according to the new approach.

Appendix XXVII

Ribosomes exist in the endoplasmic reticulum and mitochondria as small balls according to the literature. Ribosomes consist of two big units (thirty-three proteins) and small ones (forty-six proteins). The small units attract mRNA whereas the big ones attract tRNA with specific amino acids. The carrier proteins bring mRNA from the nucleus.[84]

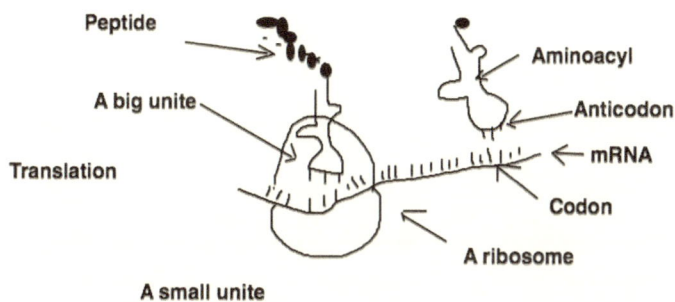

Peptide

A big unite

Translation

A small unite

Aminoacyl

Anticodon

mRNA

Codon

A ribosome

The diagram shows the contemporary dogma of protein biosynthesis that occurs between the protein subunits.

The tRNA has an anticodon of three nucleotides and specific amino acids. The codons in mRNA attract specific anticodons of tRNA with their specific amino acids. The tRNA with specific amino acid attracts another specific tRNA with specific amino acid forming polypeptides in a big unit according to the central dogma. There are many videos about protein biosynthesis on YouTube.[85, 86, 87]

The above diagram is a typical example of how protein biosynthesis is depicted in the contemporary literature. The mRNAs come between the protein subunits of ribosomes by an unclear mechanism. The newly synthesized polypeptides also get out somewhere also by an unclear mechanism. The contemporary literature does not offer a logical explanation of protein and nucleic acid biosynthesis due to dogmatic approaches to eukaryotic cells. Dr. Joseph Goebbels explained how dogma got power: "A lie told once remains a lie but a lie told a thousand times becomes truth." Contemporary biology ignores facts that are not thoroughly examined, explained, and put into a broader picture of eukaryotic cells. A good example of the confusion is in the below picture of ribosomes with this comment about them.

> Clusters of ribosomes may sit on a mRNA and make proteins, each making a strand of polypeptides. These clusters are called polyribosomes. When they are free in the cytoplasm, they are called free polyribosomes (linked by the mRNA). Or, they may bind to rough endoplasmic reticulum.

The sentences about ribosomes belong to the website that explains protein biosynthesis.[88]

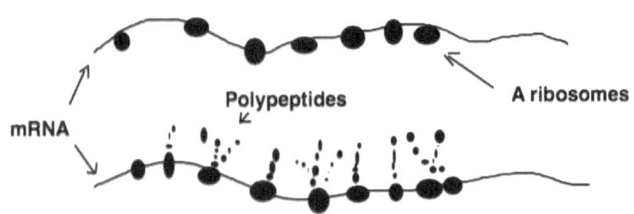

A MORPHOLOGIST VIEW OF TRANSLATION

The drawing is made according to the electronic image of ribosomes and the explanation of the translation of the author of the same website.[88]

Polypeptides in the picture are big, almost like ribosomes, and they get out of ribosomes without any questions about their sizes and origin. The same author made the new scheme to explain the formation of proteins.

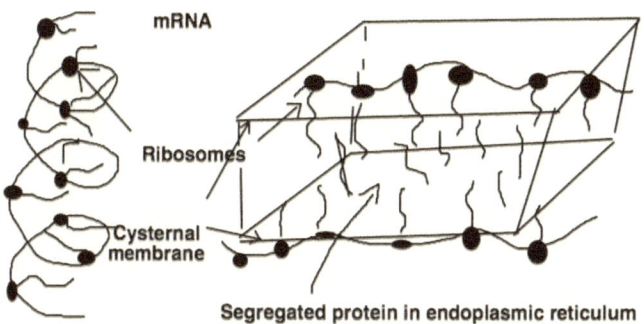

The scheme is based on the scheme of the same author.[88]

The scheme of the formation of the proteins does not explain the formation of proteins from polypeptides, the transport of new proteins, and their assembly in different organelles. Proteins float between the cisternal membranes of the endoplasmic reticulum. The unknown phenomenon somehow form organelles from different proteins.

Appendix XXVIII

According to Encyclopedia Britannica,[88] mitosis consists of these phases:

- interphase (duplication of chromosomes and their thickening, division of centrosome body)
- prophase (centrosomes move apart, an appearance of the mitotic spindle and disappearance of nucleus membrane)
- metaphase (doubled chromosomes attach the mitotic spindle by centromeres and their lining up at midcell)
- anaphase (half chromosomes move to a pole, half to the other one)
- telophase (mother cells divide into two cells)

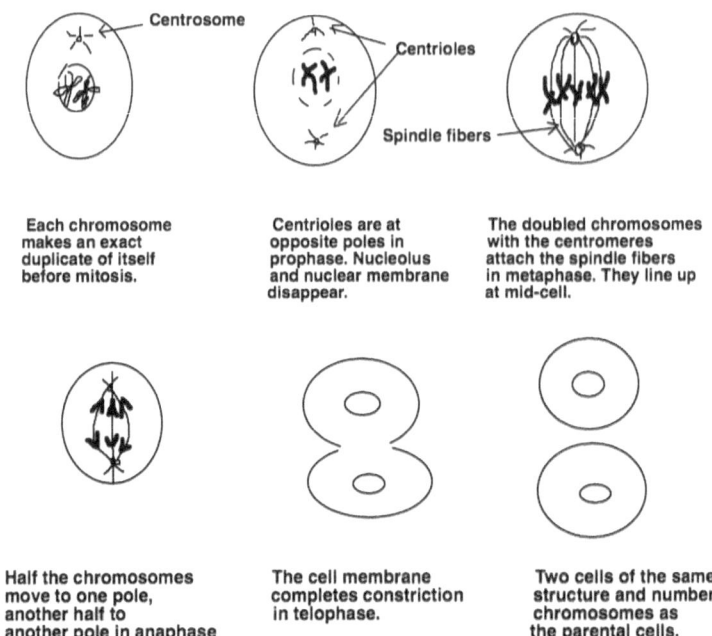

| Each chromosome makes an exact duplicate of itself before mitosis. | Centrioles are at opposite poles in prophase. Nucleolus and nuclear membrane disappear. | The doubled chromosomes with the centromeres attach the spindle fibers in metaphase. They line up at mid-cell. |

| Half the chromosomes move to one pole, another half to another pole in anaphase | The cell membrane completes constriction in telophase. | Two cells of the same structure and number chromosomes as the parental cells. |

THE DRAWING ACCORDING TO THE DRAWING OF ENCYCLOPEDIA BRITANNICA

Daughter and mother cells are very similar but never the same due to mutations in nucleic acids before mitosis. Interphase consists of three phases: G1 (duplication of a cell except for chromosomes), S (duplication of nucleic acid), and G2 (cells check for errors in chromosomes).

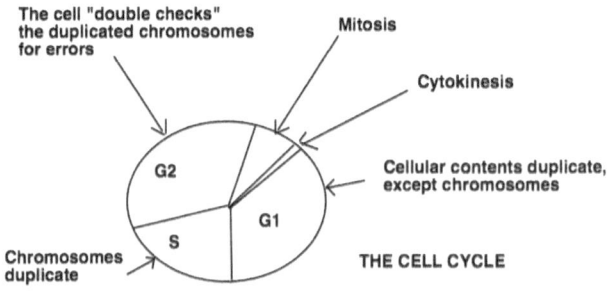

Diagram showing contemporary understanding of mitosis

The division of eukaryotic cells occurs in four phases according to the contemporary literature.

- Prophase: DNA thickens into chromosomes. A chromosome has two chromatids and a centromere. The nuclear envelope breaks down and disappears.
- Metaphase: the spindle is formed. The spindle strings attach chromosomes for centromeres. Chromosomes travel and arrange at the equatorial plate.
- Anaphase: the centromeres and sister chromatids split, separate, and move to opposite poles.
- Telophase (Cytokinesis): the chromosomes arrive at opposite poles and strings of the mitotic spindle disappear. The chromosomes expand in chromatin, and the cytoplasm divides. The nuclear membrane reappears, and the mother cell divides into two daughter cells.

The contemporary view of mitosis does not explain the logic and nature of the phases and divisions of cells because scientists do not understand the existence of the complex cylinders with their structures in eukaryotic cells.

DNA strands are in the strings of the complex cylinders and separated at the ends of the growing parts of the strings due to biosynthesis of nucleic acid and protein synthesis. DNA strands undergo uncoiling, separation, and transcription without any logical explanations of the phenomena according to the literature. This irrational approach described in the central dogma is accepted by contemporary literature. The authors of the irrational approach do not try to explain this approach because they are afraid to think.

Appendix XXIX

One hair consists of two extended loops of filaments attracted and coiled together into the octamer according to the literature. One loop of a filament consists of the four protein spirals or the tetramer, the same as a filament. The four protein spirals can be divided into the two protein spirals or the two dimers, which consist of two different protein spirals.

The balls of chromosomes are investigated and described in the literature as separate histones that do not have any connections with filaments and chromosomes because the contemporary view does not explain the structure of chromosomes by filaments. In other words, histones of chromosomes are not seen as the remains of broken filaments according to the contemporary view.[89] The histones cannot be separated from the strings because they are parts of the coiled strings according to the new view.

Appendix XXX

Meiosis occurs in sexually reproducing eukaryotes including plants, fungi, and animals.[90, 91] It occurs by the two cycles of cells divisions producing four daughter cells with the half (haploid) number of chromosomes of their mother cell. The two meiotic divisions are called meiosis I and meiosis II.

Any cell has homologous chromosomes. A pair of homologous chromosomes has one paternal and one maternal chromosome that pair up with each other during meiosis (human genomes have forty-six chromosomes with twenty-three pairs of homologous chromosomes). Homologous chromosomes have the same locations of the same genes.

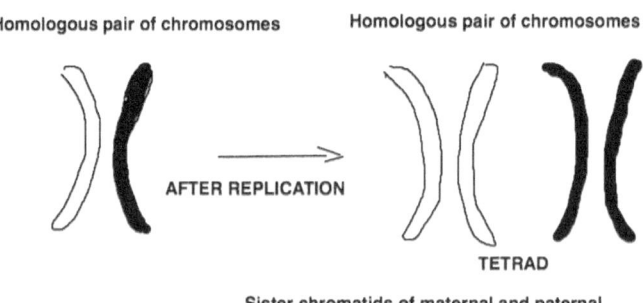

The model of duplication of chromosomes according to the literature

DNA is replicated so that any chromosome in prophase consists of two identical chromatids attached at a centromere before meiosis I during interphase (S phase).

The chromatids of homologous chromosomes pair with each other and exchange genetic material during meiosis I; this is chromosomal crossing over. Attracted chromatids stay attached at the end of meiosis I.

During meiosis II, the two daughter cells divide into four daughter cells with the haploid number of chromosomes. During meiosis II, sister chromatids detach from one another so any daughter cell gets the haploid number of chromatids (chromosomes). These daughter cells with the haploid number of chromosomes mature into gametes, spores, pollen, and other reproductive cells.

Meiosis has many different diagrams that do not explain an essence of how and why the sex cells get the haploid number of chromosomes.

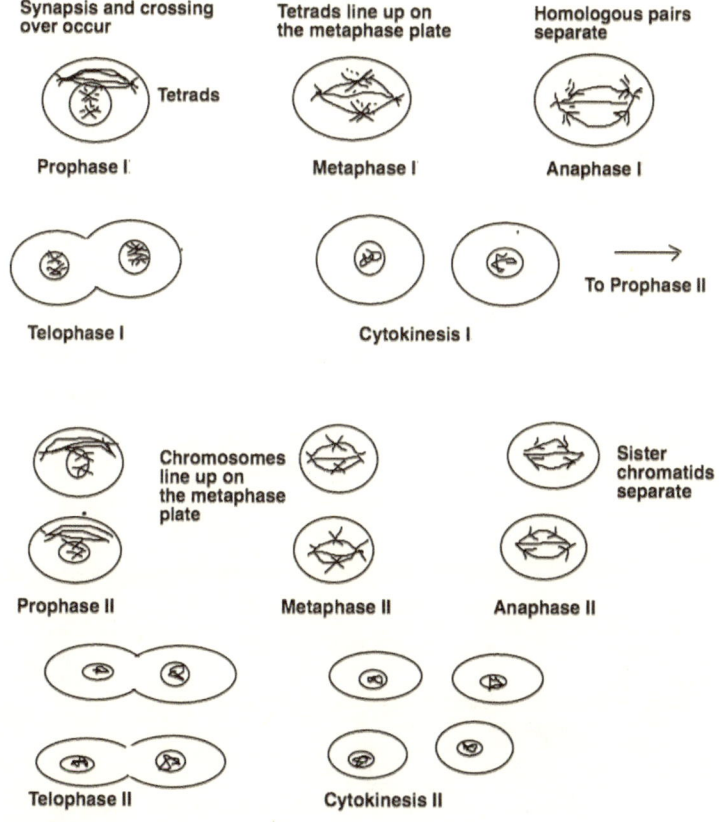

Contemporary literature does not explain clearly the occurrence and disappearing of the tetrads.[92, 93]

Appendix XXXI

Activation of Specific Genes according to the Contemporary Literature

Transcription is regulated by different mechanisms in prokaryotic and eukaryotic cells according to the contemporary literature.[94, 95] Prokaryotic cells have DNA as promoter sequences that bind RNA polymerase, operator sequences that bind the repressor protein and genes that produce specific proteins. Eukaryotic cells have DNA as promoter sequences that bind RNA polymerase, enhancer sequences that bind transcription factors, and genes that produce specific proteins. Eukaryotic cells do not have the operator sequences that bind the repressor proteins.

Prokaryotic cells have positive control when repressor proteins do not attract specific molecules (e.g., lactose molecules) due to a lack of specific molecules, and repressor proteins bind the operator sequences. Repressor proteins without specific molecules (lactose) do not allow binding of RNA polymerase for the promoter sequences to transcribe mRNA with the genes that produce the specific proteins (lactase enzymes). Repressor proteins with specific molecules have conformational changes that do not allow binding repressor proteins for the operator sequences in the case of extra specific molecules. RNA polymerase binds the promoter sequences without repressor proteins and transcription of the specific genes as mRNA occurs.

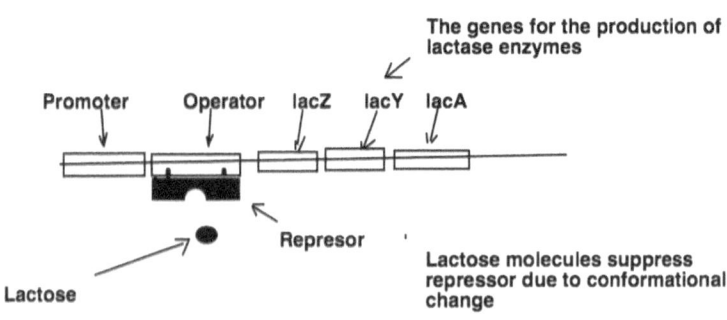

The genes for the production of lactase enzymes

Promoter   Operator   lacZ   lacY   lacA

Represor

Lactose

Lactose molecules suppress repressor due to conformational change

POSITIVE CONTROL

Prokaryotic cells have negative control when repressor proteins with specific molecules bind the operator sequences and do not allow RNA polymerase to bind the promoter sequences. When specific molecules (example, tryptophan) do not exist in the environment, molecules do not exist also in repressor proteins. The repressor proteins without the specific molecules (tryptophan) have conformational changes that do not allow binding the repressor proteins for the operator sequences. The operator sequences without the repressor proteins allow binding RNA polymerase for the promoter sequences. RNA polymerase transcribes the specific genes for the production of proteins (enzymes) for the production of the specific molecules.

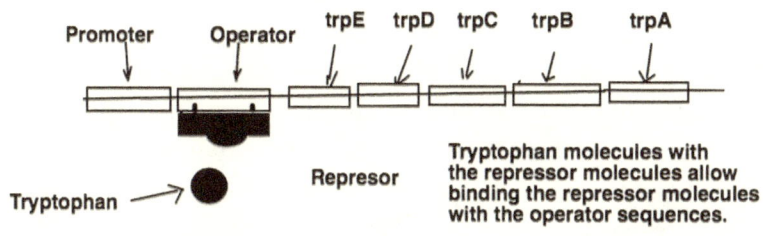

NEGATIVE CONTROL

Eukaryotic cells do not have the operator sequences and positive and negative controls of specific genes. They have transcription factors produced by regulatory genes. RNA polymerase cannot bind for the promoter sequences without transcription factors. The transcription factors bind for the promoter and enhancer sequences. Transcription factors and promoter sequences attract RNA polymerase. The complex of promoter sequences, transcription factors, and RNA polymerase attract other attraction factors attracted by the enhancer sequences. They bend back DNA together with transcription factors and enhancer sequences forming the new complex. The new complex activates RNA polymerase that transcribes the specific genes.

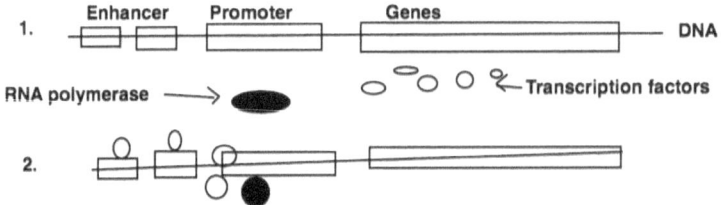

The transcription factors bind for the promoter sequences and attract RNA polymerase

3.

The transcription factors bind also for the enhancer. They bend DNA and attract RNA polymerase. The transcription factors with RNA polymerase activate it.

The new approach to regulatory genes is different from the approach described in the literature. The specific molecules at prokaryotic and eukaryotic cells change attractions among specific genes and polypeptides. Specific molecules with other ions choose polypeptides that free specific genes for decomposition or synthesis of molecules. The mechanism is the same for prokaryotic and eukaryotic cells without the complex explanations of the literature, which cannot find logical answers to repressor proteins with off and on functions and the bending mechanism with activation of RNA polymerase.

Appendix XXXII

According to E. Rosenberg and I. Zilber-Rosenberg in "The Hologenome Concept: Human, Animal and Plant Microbiota," the origin of eukaryotic cells is explained by a different hypothesis.[100]

> Since eukaryotes perform the same basic biochemical reactions as prokaryotes, i.e., the unity of biochemistry, it is highly likely eukaryotes descended from prokaryotes. For example, of the approximately 22,000 human genes that code for proteins (Stein 2004), 60 % of them are homologs of prokaryotic genes (Domazet-Loso and Tautz 2008), primarily those involved in intermediate metabolism. Many scientists assumed that eukaryotes evolved from prokaryotes through the familiar

process of mutation and natural selection. However, Lynn Margulis argued that a number of parts of the eukaryote cell were acquired in a radically different way, namely by the fusion of separate bacterial species (Sagan 1967). Many studies have bolstered this once-controversial endosymbiont hypothesis. Let us first consider the mitochondrion, a fundamental organelle of all eukaryotic cells. Mitochondria resemble bacteria in many ways. Both are surrounded by a double layer membrane, and the structural arrangement of the inner and outer membrane of mitochondria is similar to Gram-negative bacteria. Mitochondria and bacteria can use oxygen to generate chemical energy, in the form of adenosine triphosphate (ATP) molecules. The ribosome coded by mitochondrial DNA is similar to those from bacteria in size and structure. Mitochondria have their own bacterial-like DNA, which they duplicate when they divide, similarly to bacteria, into new mitochondria. Based on DNA analyses, mitochondria are closely related to the intracellular parasite Rickettsia prowazekii (Andersson et al. 1998), the causative agent of epidemic typhus. Using similar arguments, it is now generally accepted that chloroplasts of photosynthetic algae and plants arose through incorporation into cells of symbiotic cyanobacteria (Martin et al. 2002). In fact, the data indicate that chloroplasts were acquired through multiple primary endosymbiotic events (Fagan and Hastings 2002) that occurred long after the acquisition of mitochondria (Perasso 1989). If one accepts the bacterial origin of mitochondria and chloroplasts, then the eukaryote cell, itself, is a holobiont. One of the fundamental differences between eukaryotes and prokaryotes is the presence of a membrane-bound nucleus in eukaryotes. Because the nucleus lacks an obvious homologue or precursor among prokaryotes, ideas about its evolutionary origin are diverse and highly speculative. One hypothesis is that a prokaryotic cell membrane formed an invagination that enclosed the DNA in a primitive prokaryote (probably an Archaea) and this membrane-bound DNA evolved into the nucleus (Jékely 2003). Another hypothesis is endosymbiosis of an archaebacterium within an eubacterium, with the archaebacterium becoming a nucleus (Martin 2005). The reason for assuming that the archaebacterium became the nucleus is that the molecular machinery involved in information storage and retrieval in eukaryotes is more similar to archaebacterial counterparts than to eubacterial counterparts (Reeve 2003). On the other hand, eukaryotic genes involved in metabolic and biosynthetic pathways reflect a eubacterial ancestry (Simonson et al. 2005).

The new approach contradicts the above approach in many ways.

1. Prokaryotic and eukaryotic cells had the same predecessors. Eukaryotic cells had not descended from prokaryotic cells because prokaryotic cells have filaments whereas eukaryotic cells have the complex cylinders.
2. The mitochondrion does not resemble prokaryotic cells because the complex cylinder builds mitochondrion (the chloroplast) and a prokaryotic cell consists of many filaments.
3. Eukaryotic cells had not been born by fusion of prokaryotic cells because archaic prokaryotic cells had been born by fusion of filaments whereas archaic eukaryotic cells had been born by fusion of archaic complex cylinders.

Appendix XXXIII

According to E. Rosenberg and I. Zilber-Rosenberg in "The Hologenome Concept: Human, Animal and Plant Microbiota," the origin of the first multicellular eukaryotic organism is explained by a different hypothesis.[100]

The origin of the first multicellular eukaryotic organism has been a topic of intense debate in biology, and many hypotheses have been put forth to explain this evolutionary milestone (Grosberg and Strathmann 2007). It is reasonable to hypothesize that early eukaryotic cells, formed by the fusion of two or more prokaryotes, had the genetic information that would allow for cell-to-cell interactions and the formation of multicellular structures. Support for this hypothesis comes from the discovery that morphogenesis of a choanoflagellate (one of closest living relatives of animals) is induced by bacteria in the Bacteroidetes phylum (Alegado 2012). Further, it was shown that the inducing factor is a bacterially produced sulfonolipid. This study provides another example of how bacteria may have contributed to the evolution of animals. The relative ease at which unicellular organisms can evolve into multicellularity is supported by the fact that multicellularity has evolved independently dozens of times in the history of Earth, for example at least once for plants, once for animals, once for brown algae, and several times for fungi, slime molds, and red algae (Bonner 1998). The earliest animal that still exists is the sponge. What can the sponge tell us about the early evolution of animals? Costerton et al. (1995) have compared modern sponges to biofilms because both lack tissues and organs, but are composed of a three-dimensional matrix that allows for the flow of water, nutrients, metabolites, and oxygen. Modern sponges are well known for containing large complex microbial symbiotic communities. More than half of the biomass of some sponges is bacteria (Taylor et al. 2007). The fossil record of sponges demonstrates their ancient association with bacteria, further indicating that prokaryotic symbionts were essential components of animals from their very beginning. Interestingly, some present sponge symbionts produce proteins that have domains that have cell-attachment activity (Siegl et al. 2010). One could speculate that similar bacterial proteins were involved in providing the "glue" for the construction of the first multicellular eukaryotes. Some evidence exists for specific genes involved in early multicellularity (Rokas 2008). It should be pointed out that not all manifestations of multicellularity are the same. For example, multicellularity in volvocine green algae likely evolved as a consequence of incomplete separation after cell division, whereas in cellular slime molds multicellularity evolved as a consequence of aggregation (Waggoner 2001).

The first multicellular eukaryotic organisms had been formed by fusion of eukaryotic cells in colonies according to the new approach. The new approach contradicts the same authors' hypothesis.

In light of the available information, we propose that animal and plant cells arose from prokaryotic organisms by fusion, aggregated into multicellular complexes, initially using prokaryotic genetic information, and differentiated into animals and plants, always in close association with microorganisms. During evolution, which is discussed in Chap. 8, animals and plants acquired additional structures and functions either by changing their DNA or by acquiring new symbionts. Good examples of the latter are ruminants (Dehority 2003) and termites (Brune 2011), which evolved the ability to utilize cellulose as a nutrient by incorporating cellulose-decomposing microorganisms, thereby avoiding the very slow process of evolving novel efficient enzyme systems and regulatory elements by themselves.
100

The above hypothesis is also a good example of the dogmatic approach because it speculates about different hypotheses.

Appendix XXXIV

The literature explains the entrance of the viral nucleic acids into prokaryotic cells by the bacteriophage injection.

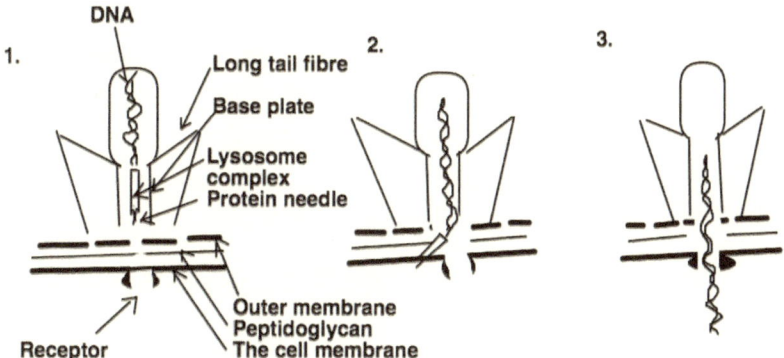

Myovirus bacteriophages inject their genetic material into the cell by "a hypodermic syringe-like motion" with the help of ATP present in the tail.[96]

Appendix XXXV

Viruses replicate in seven stages according to the literature.[97]

1. Adsorption—the virus is attracted by the receptor of the cells membrane.
2. Entry—the cell membrane invaginates with the virus and protects against antibodies.
3. Uncoating—the virus goes to a lysosome by an unknown way. The lysosome's enzymes strip off the viral proteins and release viral nucleic acids.
4. Transcription/mRNA production—the different viral nucleic acids by the enzymes transfer in mRNA and get in ribosomes of the host without a logical explanation how the viral mRNA gets there.
5. Synthesis of virus components—the viral mRNA synthesize two groups of the viral proteins: structural—the proteins which

make up the virus particle are manufactured and assembled—and nonstructural—not found in particle, mainly enzymes for virus genome replication.

6. Virion assembly—the newly synthesized genome (nucleic acid) and proteins are assembled to form new virus particles. This occurs in the cell membrane, cytoplasm, and nucleus.

7. Release (liberation stage)—the viruses are released by either sudden rupture of the cell, or gradual extrusion (budding) of enveloped viruses through the cell membrane.

Appendix XXXVI

In "Hepatitis B replication" (2007), Juergen Beck and Michael Nassal write,[98]

(1) Infectious virions contain in their inner icosahedral core the genome as a partially double-stranded, circular but not covalently closed DNA of about 3.2 kb in length (relaxed circular, or RC-DNA); (2) upon infection, the RC-DNA is converted, inside the host cell nucleus, into a plasmid-like covalently closed circular DNA (cccDNA); (3) from the cccDNA, several genomic and subgenomic RNAs are transcribed by cellular RNA polymerase II; of these, the pregenomic RNA (pgRNA) is selectively packaged into progeny capsids and is reverse transcribed by the co-packaged P protein into new RC-DNA genomes. Matured RC-DNA containing-but not immature RNA containing-nucleocapsids can be used for intracellular cccDNA amplification, or be enveloped and released from the cell as progeny virions. Below we discuss these genome conversions, with emphasis on the reverse transcription step, and particularly its unique initiation mechanism.[98]

The authors of the same article use the scheme that explains replication of hepatitis B.

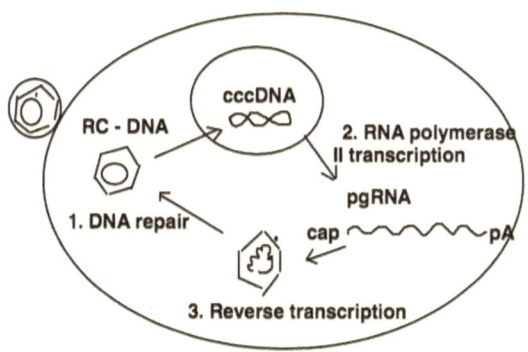

Drawing according to the scheme of Jurgen Beck and Michael Nassal.[98]

Enveloped virus infects the cell and releases RC-DNA (1). RC-DNA is transported to the nucleus where it forms cccDNA after the repair. RNA polymerase II transcribes cccDNA and produce pgRNA (2). The pgRNA is encapsulated with P protein and reverse transcribed inside the nucleocapsid. (+)-DNA synthesis from (-) – DNA generates new RC-DNA. The new cycles produce amplification of cccDNA.

In "Hepadnavirus Genome Replication and Persistence" (2015), Jianming Hu and Christoph Seeger wrote,[99]

> The exact mechanism for cccDNA formation is still obscure and represents a major gap in our knowledge of the hepadnavirus life cycle. Infection of ducklings with the duck hepatitis B virus (DHBV) revealed that cccDNA formation can occur within a few hours after infection (Tagawa et al. 1986).

The same authors wrote,

> A critical, but still not-well-understood step in the replication cycle of hepadnaviruses is the sequestration of pgRNA from the translation machinery, which coincides with or is followed by pgRNA packaging into core particles (Fig. 1).

Appendix XXXVII

According to E. Rosenberg and I. Zilber-Rosenberg ("The Hologenome Concept: Human, Animal and Plant Microbiota,"[100]), the origin of viruses is explained by three main hypotheses.

(1) Viruses arose from intracellular genetic elements that gained the ability to move between cells. According to this hypothesis, viruses originated through a progressive process. Mobile genetic elements, pieces of genetic material capable of moving within a genome, gained the ability to exit one cell and enter another.

(2) Viruses are remnants of cellular organisms. In contrast to the progressive process of hypothesis 1, viruses may have originated via a regressive, or reductive, process. Microbiologists generally agree that certain bacteria that are obligate intracellular parasites, like Chlamydia and Rickettsia species, evolved from free-living ancestors. It follows, then, that existing viruses may have evolved from more complex, possibly free-living organisms that lost genetic information over time, as they adopted a parasitic approach to replication.

(3) Viruses predate or coevolved with their current cellular hosts. It should be pointed out that these hypotheses are not mutually exclusive. To distinguish between these hypotheses it would be useful to date the origin of viruses. Unfortunately, it has not yet been possible to detect ancient viruses in fossils (Emerman and Malik 2010). However, there is strong circumstantial evidence that viruses emerged very early in the evolution of life, before the separation of prokaryotes into two domains: Eubacteria and Archaebacteria. The evidence is the striking structural similarities between viruses that infect organisms belonging to the different domains of life. For example, archaeal and bacterial tailed phages show remarkable morphological similarity (Zillig et al. 1996), and the Sulfolobus islandicus rod-shaped archaeal virus SIRV shows structural and mechanistic similarities to eukaryotic poxviruses (Filée et al. 2003). If the viruses arose after the separation into the three domains of life.

Viruses have the same origin as prokaryotic and eukaryotic cells according to the new approach. The new approach is closest to the third hypothesis that says viruses emerged very early in the evolution of life before the separation of prokaryotes into two domains—eubacteria and archaebacteria.

Viruses have the same origin as prokaryotic and eukaryotic cells according to the new approach. They stopped their development at the stage of the archaic strings.

Appendix XXXVIII

Mutation is a change of nucleotide sequences in a gene. Mutations can be due to spontaneous mutations (molecular decay), error-prone replication bypass of naturally occurring DNA damage, errors introduced during DNA repair and induced mutations by mutagens (Wikipedia).[101]

- Spontaneous mutations (molecular decay) take place on the molecular level.
  - o The repositioning of a hydrogen atom changes a base. The result is an incorrect base pairing during replication (tautomerism).
  - o Loss of a purine base (A or G) from an apurinic site (depurination).
  - o Replacement of an amino group with a keto one due to hydrolysis changes a normal base to an atypical base (deamination).
  - o Denaturation of the new strand from the template during replication followed by renaturation in a different spot ("slipping"). This can lead to insertion or deletions (Slipped-strand mispairing).
- Mutations due to error-prone replication bypass of naturally occurring DNA damage. These mutations take place after breaking a single strand of nucleic acid (template) and during repair of its broken strands. Some nucleotides are substituted, lost, or added.

- Errors introduced during DNA repair. Mutations take place after breaking of double strands and during their repairing. Some nucleotides are lost or some are added during their connection.
- Mutations caused by mutagens
  o Chemicals
  o Radiation
  o Ultraviolet radiation.
- Environmental conditions as lack of amino acids, low pH, increased extreme temperatures, UV, mutagens and other can cause point mutations during nucleic acid synthesis. Point mutations are single-base substitutions.

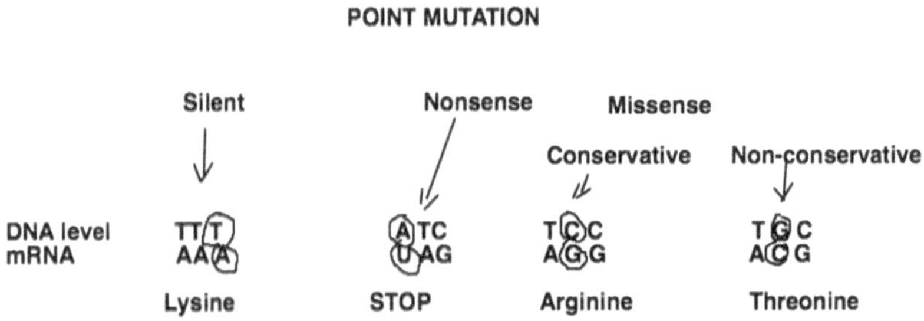

POINT MUTATION

	Silent	Nonsense	Missense	
			Conservative	Non-conservative
DNA level	TT T	A TC	T C C	T G C
mRNA	AA A	U AG	A G G	A C G
	Lysine	STOP	Arginine	Threonine

The above diagram according to Wikipedia shows the different point mutations.

- Silent mutations take place when a nucleotide is changed, but despite this, proteins are not changed during translation. Old and new codon attracts the same amino acid.
- Nonsense mutations take place also when a nucleotide is changed. New codon does not attract amino acids and therefore, this codon is an interruption codon.
- Missense mutations take place also when a nucleotide is changed. A missense mutation changes a codon so that a different protein is produced. There are two types of missense mutation.
  o Conservative mutation is when the property of new amino acid is the same as the old one (e.g., hydrophobic,

hydrophilic and so on). Most proteins withstand one or two point mutations without changes of their functions.

o   Nonconservative mutation occurs when the property of new amino acid is different from the old one. Protein is changed and loses its function that causes some diseases such as sickle cell anemia.

There are four other types of mutations except point mutation.

Oncogenes

A proto-oncogene is a normal gene that can become an oncogene due to mutation. An oncogene is a gene that has the potential to cause cancer. Oncogenes are mutated proto-oncogenes in tumor cells. Most normal cells undergo a programmed form of rapid cell death when some genes mutate but not with proto-oncogenes. Oncogenes produce oncoproteins that help to regulate cell growth and differentiation in three ways.

I.   A change in the protein structure: an increase in enzyme activity or loss of regulation
II. An increase in the amount of certain protein due to gene duplication: an increase in protein expression or protein stability

A chromosomal translocation (the type of chromosomal abnormality) with two types: translocation events relocate a proto-oncogene to a new chromosomal site that leads to a higher expression; translocation events that lead to fusion of a proto-oncogene to a second gene (the new protein has a fusion protein with increased oncogenic activity). Examples: the expression of the hybrid protein in dividing stem cell in the bone marrow leads to adult leukemia. Philadelphia chromosome (fusion of parts of DNA of chromosome 22 (BCR gene) and 9 (ABL gene). Fused BCR-ABL displays high tyrosine kinases activities that lead cells into uncontrollable growths.

The contemporary approach to understanding oncogenesis contradicts the new approach of the pathogenesis of malign cells. The contemporary approach observes mainly pathogenesis of cancer in causative relationships of proto-oncogenes, oncogenes, and oncoproteins whereas the new approach observes pathogenesis of cancer in causative relationships of known phenomena of eukaryotic cells. The drawing shows an example of the contemporary approach.

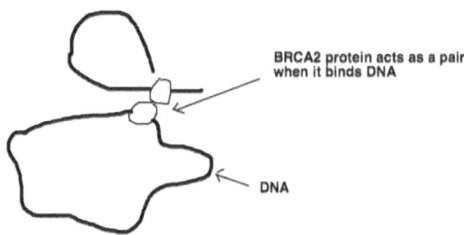

The drawing (according to the photography of Drs. Sarah Compton and Jack Griffith at The UNC Lineberger Comprehensive Cancer Center (UK), 2010) shows BRCA2 acts as a pair when it binds circular DNA. Drs. Sarah Compton and Jack Griffith hope that purification of BRCA2 could lead to a better understanding of how the protein works and how BRCA2 sequence mutations cause cancer.[102] The understanding pathogenesis of cancer without understanding the different phenomena of eukaryotic cells in causative relationships is impossible according to the new approach.

Appendix XXXIX

Taxonomy is the science of the classification of organisms. The aim of taxonomy is to identify organisms and to show their evolutionary relationships. Prokaryotic cells are classified according Bergey's *Manual of Bacteriology*. The latest edition was published in 1994. Below are some specific characteristics of prokaryotic cells used for their classification.

- Shape (morphological characteristics)
- Differential staining (gram stain)
- Nutritional pattern
- Relationship with oxygen

- Biochemical characteristics
- Serological analysis
- Phage typing
- DNA sequencing
- Analysis of 16s ribosomal sequences
- Analysis of fatty acids found in bacterial membrane
- Others

The definition a bacterial species does not exist according to the literature.

In "In search of a bacterial species definition" (1996), Edgardo Moreno[103] wrote,

> For over a century, the biological species concept in the first instance, and following this, the taxonomic species concept were the major theoretical postulates orienting bacteriological research. In the 19th century and the first half of the 20th century bacteriology was primarily confined within the limits of medical research (Smith and Martin 1949). During this time many different pathogenic bacteria were described, and sets of key phenotype characters were established to identify members of the same. It was mainly in the second half of this century that research in bacteria devoid of medical importance began to participate in the microbiologist game. Owing to the incursion of molecular biology during the last decade, the bacterial world began to develop an evolutionary historical dimension which has greatly contributed to correcting some of the distorted perceptions "not in tune with the position of microorganisms in the natural order of things" (Woese 1987). The methods used for the definition of biological and taxonomic species concepts lack sufficient evolutionary perspective and, hence provide an inappropriate guide to the origins and products of evolutionary diversification

(O'Hara 1993, Woese 1994). This has led to a call for the replacement of the biological-taxonomic concept by the phylogenetic species concept. However, as stated before, the methods used for constructing bacterial phylogenies do not in themselves possess sufficient resolution to distinguish those terminal taxons we call species; therefore a polyphasic phylogenetic-taxonomic approach was suggested (Stackebrandt and Goebel 1994). Although this latter proposition is closer to a practical definition of bacterial. species (or nomenspecies), it is vague with respect to the biological properties of the bacterial organisms. Furthermore, it does not take into consideration the microbiologist point of view, which I consider essential for a practical definition of bacterial species. As a scientist I am obligated to observe, measure, analyse and describe phenomena; however as a human being I may have preconceived notions of the object I am devoted to studying; this I can not neglect but if I recognise its existence, I may be able to control it.

In "What are bacterial species?" (2002), F. M. Cohan[104] wrote,

Bacterial systematics has not yet reached a consensus for defining the fundamental unit of biological diversity, the species. The past half-century of bacterial systematics has been characterized by improvements in methods for demarcating species as phenotypic and genetic clusters, but species demarcation has not been guided by a theory-based concept of species. Eukaryote systematists have developed a universal concept of species: A species is a group of organisms whose divergence is capped by a force of cohesion; divergence between different species is irreversible; and different species are ecologically distinct. In the case of bacteria, these universal properties are held not by the named species of systematics but by ecotypes. These are populations of organisms occupying the same ecological niche, whose divergence is purged recurrently by natural selection. These ecotypes can be discovered by several universal sequence-based approaches. These molecular methods suggest that a typical named species contains many ecotypes, each with the universal attributes of species. A named bacterial species is thus more like a genus than a species.

Despite the fact that the definition of prokaryotic species does not exist, people reclassify the species within the genus. In "Phylogenomic Analyses and Reclassification of Species within the Genus Tsukamurella: Insights to Species Definition in the Post-genomic Era" (2016), J. L. Teng,[1] Y. Tang,[2] Y Huang,[2] F. B. Guo,[3] W. Wei,[4] J. H. Chen,[2] S. S. Wong,[1] S. K. Lau,[5] and P. C. Woo[5,105] wrote,

Owing to the highly similar phenotypic profiles, protein spectra and 16S rRNA gene sequences observed between three pairs of Tsukamurella species (Tsukamurella pulmonis/Tsukamurella spongiae, Tsukamurella tyrosinosolvens/Tsukamurella carboxy-divorans, and Tsukamurella pseudospumae/Tsukamurella sunchonensis), we hypothesize that and the six Tsukamurella species may have been misclassified and that there may only be three Tsukamurella species. In this study, we characterized the type strains of these six Tsukamurella species by tradition DNA-DNA hybridization (DDH) and "digital DDH" after genome sequencing to determine their exact taxonomic positions. Traditional DDH showed 81.2 ± 0.6% to 99.7 ± 1.0% DNA-DNA relatedness between the two Tsukamurella species in each of the three pairs, which was above the threshold for same species designation. Digital DDH" based on Genome-To-Genome Distance Calculator and Average Nucleotide Identity for the three pairs also showed similarity results in the range of 82.3-92.9 and 98.1-99.1%, respectively, in line with results of traditional DDH. Based on these evidence and according to Rules 23a and 42 of the Bacteriological Code, we propose that T. spongiae Olson et al. 2007, should be reclassified as a later heterotypic synonym of T. pulmonis Yassin et al. 1996, T. carboxydivorans Park et al. 2009, as a later heterotypic synonym of T. tyrosinosolvens Yassin et al. 1997, and T. sunchonensis Seong et al. 2008 as a later heterotypic synonym of T. pseudospumae Nam et al. 2004. With the advancement of genome sequencing technologies, classification of bacterial species can be readily achieved by "digital DDH" than traditional DDH.

Appendix XL

E. Rosenberg and I. Zilber-Rosenberg explain the origin of life in "The Hologenome Concept: Human, Animal and Plant Microbiota."

The origin of life is one of the most challenging problems in biology. The best evidence is that prokaryotic life (microorganisms lacking a membrane-bound nucleus) first appeared on Earth about 3.8 billion years ago (Zimmer 2009). The abiogenesis theory of the origin of life begins with the experimentally supported concept that simple organic compounds, such as amino acids, organic acids, purines, pyrimidines, and simple sugars, could have been produced spontaneously in the earth's primitive reducing atmosphere (Oró and Kamat 1961), containing CH_4, H_2, H_2O, N_2, and NH_3 (Schaefer and Fegley 2010). Further, these molecules could have been concentrated into a "prebiotic organic soup" in evaporation ponds. Whatever the earliest events on the road to the first living cell were, it is clear that at some point the large biological molecules found in modern cells must have emerged. Considerable debate in origin-of-life studies has revolved also around which of the fundamental macromolecules came first—proteins, DNA, or RNA. To gain insight into this question it is useful to consider the functions performed by each of these polymers in existing organisms.

The new approach is not in a contradiction with this approach, although the new approach explains evolution of prokaryotic cells by evolution of polypeptides that build protein spirals, which build strings, which build filaments, which build prokaryotic cells and their organelles.

Appendix XLI

Methods of the New Taxonomy

The methods are built according to the new taxonomy of prokaryotic cells and the different needs of industrial and medical microbiology. The new methods of the new classification will be used for identification of different prokaryotic cells, storage of different prokaryotic species, and designing prokaryotic ecosystems for the different needs of industry, agriculture, and medicine.[100]

Different environments will be investigated for different reasons.

- finding pathogens and their neutralization by different means[111]
- finding toxic substances, their influence on life, and their neutralization by different means including decomposition by designed prokaryotic ecosystems[106, 107, 108]
- decomposition of different waste into useful organic compounds by designed prokaryotic ecosystems[109]
- enrichment of ground with organic compounds and prokaryotic ecosystems designed for agricultural needs
- production of organic compounds by designed prokaryotic ecosystems for medical and industrial needs
- other reasons[112]

The existing taxonomy of prokaryotic cells with its methods is not practical for the above purposes because it does not have the tools for understanding the metabolism of different prokaryotic cells and metabolic connections among different prokaryotic cells in different prokaryotic ecosystems. The new taxonomy, the new methods, and the new approach to the metabolism of prokaryotic cells can provide the following.

Specimens

Reasons for investigation of environments must be known before investigation of their physical, chemical, and biological properties due to finding efficient solutions for existing problems. Some examples of reasons for investigation of environments are these.

- different waste with toxic compounds, their influence on life, and their neutralization
- organisms that produce diseases of plants, animals, humans, and their neutralization
- improvement of agriculture by different prokaryotic ecosystems

Environments of future specimens are investigated for the presence of air, different pH, salinity, the sun, and different temperatures. Data about the environment and representative specimens are sent to chemical lab for chemical analysis and microbiology labs for identification of prokaryotic species. The chemical contents of the specimens determine the presence of specific prokaryotic cells because prokaryotic cells need specific compounds for metabolism.

Specimens for microbiology labs are processed according to their consistency. If they are solid or semisolid, they are ground, distilled, put in tubes under aerobic and anaerobic conditions. They are also put in Brewer's medium for enhanced growth of anaerobic prokaryotic cells.[113] Anaerobic cabinets are used for grinding specimens to preserve anaerobic prokaryotic cells.

Media for Microbiology Investigations

Many prokaryotic cells do not grow in artificial media. The problem is partially overcome by the introduction of natural media for examination of their characteristics according to practical classifications. Natural media are made of representative specimens of environments where prokaryotic cells live. Despite this, unknown prokaryotic cells must be grown in standard media to examine the powers of metabolism. The standard medium is artificial. If prokaryotic cells do not grow in the standard medium, only the natural media with additions for determining metabolic powers are used.

Around six kilograms of the specimen is taken from the environment for preparation of the natural medium. The hard specimen is ground under aerobic conditions and put in the sterile bucket with six liters of sterile water with Na-thioglycollate. If the specimen is semisolid, it is put in a stainless steel bucket also with six liters of sterile water with Na-thioglycollate. The specimen is mixed with water and boiled for a short time. The specimen with water is filtered in another sterile bucket to get the natural medium that is clear. The natural medium of the specific volume (100ml) is distributed in thirty-six sterile bottles that contain the single coenzymes and the single metabolites of the standard concentrations. A sterile bottle contains only the natural medium. The rest of natural medium is distributed in the big bottle for the preparation of agar plates (1 liter). The natural medium is used for the preparation of the agar plates for separation of the specific organisms of the specific specimen from which is prepared the natural medium.

- 1l of natural medium
- 1 g/l of yeast extract
- 9 g/l tryptone
- 7.5g/l Agar

This formulation is used for the preparation of the standard medium for the microtiter plates: 53286 SIGMA-ALDRICH

The natural medium with the coenzymes of the standard concentration (0.01 gmol/l) and the standard medium also with the coenzymes of the standard concentration (0.01 gmol/l) are transferred into wells of microtiter plates by pipette in this order:

1. Natural medium (standard medium)
2. Natural medium (standard medium)+ NAD coenzymes
3. Natural medium (standard medium) + FMN coenzymes
4. Natural medium (standard medium) + ATP coenzymes
5. Natural medium (standard medium) + B1 coenzymes
6. Natural medium (standard medium) + folic acid coenzymes
7. Natural medium (standard medium) + biotin coenzymes
8. Natural medium (standard medium) + CoA coenzymes
9. Natural medium (standard medium) + K vitamins
10. Natural medium (standard medium) + B6 coenzymes
11. Natural medium (standard medium) + delta-aminolevulinic acids
12. Natural medium (standard medium) + FeSO4

The natural and standard media with simple metabolites of standard concentration (0.01 gmol/l) are transferred into wells of microtiter plates by pipette in this order:

1. Natural medium (standard medium) + fructose
2. Natural medium (standard medium) + (add HCl to make pH 6)
3. Natural medium (standard medium) + (add NaOH to make pH 8)
4. Natural medium (standard medium) + Na 3-Phosphoglycerate
5. Natural medium (standard medium) + Na-pyruvate
6. Natural medium (standard medium) + Na-acetate
7. Natural medium (standard medium) + Na-oxaloacetate
8. Clear medium (standard medium) + Na-acetoacetate
9. Clear medium (standard medium) + Na-propionate
10. Natural medium (standard medium) + alanine
11. Natural medium (standard medium) + glycine
12. Natural medium (standard medium) + tryptophan

The natural medium with the complex metabolites (lactose, galactose, mannose, saccharose, and urea) of standard concentration (0.01 gmol/l) and the complex metabolites (peptidoglycan, cellulose, starch, PAH, PET or MEHT, collagen, and deoxyribonucleic acid) of the standard concentration (0.01 g/l) are transferred into wells of microtiter plates by pipette in this order:

1. Natural medium (standard medium) + lactose
2. Natural medium (standard medium) + galactose
3. Natural medium (standard medium) + mannose
4. Natural medium (standard medium) + saccharose
5. Natural medium (standard medium) + peptidoglycan
6. Natural medium (standard medium) + cellulose
7. Natural medium (standard medium) + starch
8. Natural medium (standard medium) + PAH*
9. Natural medium (standard medium) + urea
10. Natural medium (standard medium) + PET**
11. Natural medium (standard medium) + collagen
12. Natural medium (standard medium) + deoxyribonucleic acid

*Polycyclic aromatic hydrocarbons (PAHs) are mutagenic, cytotoxic, and carcinogenic organic chemicals. PAHs are widely distributed in the environment because of the incomplete combustion of organic matter, emission sources, automobile exhaust, domestic matter, and other factors. Laccases can catalyze the oxidation of phenols, polyphenols, and anilines.[114]

**PET stands for polyethylene terephthalate, a plastic with mechanical, barrier, and optical properties. ISF6_0224 protein produces the conversion of mono (2-hydroxyethyl) terephthalic acid (MEHT) into PET's two environmentally benign monomers, terephthalic acid and ethylene glycol.[115]

The natural and standard medium with separately added different coenzymes and simple and complex metabolites are put into wells of three microtiter plates according to the order by a multichannel pipette. The first microtiter plate contains the natural medium and the standard medium with the coenzymes, the second contains the natural medium and the standard medium with the simple metabolites, and the third contains the natural medium and the standard with the complex metabolites. The three microtiter plates can investigate four different prokaryotic cells.

The natural medium and the standard medium are also put in tubes used for investigations of the environmental characteristics. HCl, NaOH, and NaCl are added to the tubes to make standardized characteristics. The sixteen tubes are used for an examination of a prokaryotic species. The eight tubes are for aerobic examinations and the eight tubes are for anaerobic examinations.

1	2	3	4	5	6	7	8
aerobic condition	anaerobic condition	pH 5	pH 8	60% NaCl	sun rays	9C	50C

The processing of the specimens for microbiology investigations

The technician inoculates the specimens from the tubes with distilled water and Brewer's medium into the agar plates with natural and the standard medium under anaerobic conditions. The technician streaks the agar plates by a loop also under anaerobic conditions. After this, the technician incubates the agar plates under aerobic and anaerobic conditions and the temperature of the specimen that exists in nature for one to three days.

The technician observes plates for the growth of colonies during incubation. If there are growths, the technician takes colonies by the straight wire and transfers them to the microscope slide with a drop of sterile water and to the tubes with the natural and standard medium for investigations of the characteristics according to the different environmental conditions.

1	2	3	4	5	6	7	8
aerobic conditions	anaerobic conditions	pH 5	pH 8	60% NaCl	sun rays	9C	50C

The technician investigates prokaryotic cells according to the practical classification.

1. Shape

2. Gram stain
3. Surviving under air or without air or both, different pH, high salinity, sun rays, and low and the high temperatures
4. Speed of growth in standard medium
5. Quantitative order of specific coenzymes according to different coenzymes concentrations in prokaryotic cells
6. Synthetic or decomposition group
7. Frequency of depolarization
8. enzymes that utilize complex molecules and their quantities
9. Identification according to contemporary classification

The technician stains the microscope slide with prokaryotic cells by the gram stains. The gram staining describes the shapes of prokaryotic cells and their cell walls.

No growths or growths of prokaryotic cells in tubes with the natural medium for determination of the environmental characteristic determine the environmental characteristics. They grow or do not grow under aerobic, facultative, or anaerobic conditions, under extreme pH, the high salt concentrations, the sun, and extreme temperatures.

1	2	3	4	5	6	7	8
Aerobic conditions	Anaerobic condition	pH 5	pH 8	60% NaCl	Sun rays	9C	50C

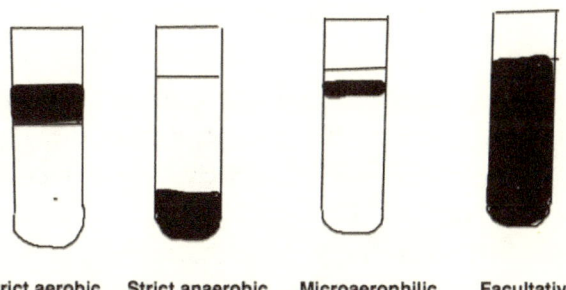

Strict aerobic   Strict anaerobic   Microaerophilic   Facultative

Prokaryotic cells live under different oxygen demands.

The technician transfers prokaryotic cells from the tubes into the wells of the microtiter plates by a straight wire.

Microtiter Plate Method

The technician transfers the same prokaryotic cells from the tubes into all wells of the microtiter plates by the straight wire. After transferring the same prokaryotic cells, the technician incubates microtiter plates under environmental temperature. If the prokaryotic cells are anaerobic, they are incubated under anaerobic conditions (the wells are covered with mineral oil). If prokaryotic cells are aerobic or facultative, they are incubated under aerobic conditions covered with sealing film.

The technician determines the optic densities of the wells of the microtiter plates before incubation (OD1) and the optic densities after incubation (OD2) by the ELISA machine, including the time before incubations (t1) and the time after incubation (t2).

The optic densities of the wells are determined every three hours. When the increased optic densities appear in all wells of the microtiter plates, the optic densities of all wells are determined (OD2) including the time (t2).

The speed of growth (S) and concentrations of different coenzymes of prokaryotic species are calculated according to the formula which are explained in chapter 15.

The quantitative relationships among metabolically connected coenzymes (ATP, B_1, folic acid, biotin, CoA, K, B_6) determine if the examined prokaryotic cells are decomposition or synthetic ones regarding metabolism of carbohydrates, fats, and proteins. If they produce more complex metabolites than simple ones, they are synthetic prokaryotic cells. Synthetic prokaryotic cells have these quantitative relationships among metabolically connected coenzymes: (ATP > B_1 > folic acid > biotin or CoA, CoA < K < ATP). If they produce fewer complex metabolites than simple ones, they are decomposition cells. Decomposition prokaryotic cells have these quantitative relationships among metabolically connected coenzymes: (ATP < B_1 < folic acid < biotin or CoA, CoA > K > ATP).

The quantitative relationships among metabolically connected coenzymes (NAD, FMN, and ATP, B_1, folic acid, biotin, CoA, K, B_6)

determine if the examined prokaryotic cells have high, medium, or low frequency of depolarization.

Prokaryotic cells with a high frequency of depolarization produce small complex metabolites if they are synthetic ones or small simple metabolites if they are decomposition, because prokaryotic cells with high frequency of depolarization do not have enough time to build big complex and big simple molecules as do prokaryotic cells with low frequency. The small complex molecules mean small molecules of polysaccharides, peptidoglycan, and fatty acids. The small simple molecules mean the big presence of products of glycolysis and Krebs cycle and the small presence of monosaccharide and disaccharide molecules.

Prokaryotic cells with high frequency of depolarization have bigger concentration of NAD and FMN coenzymes than do other coenzymes (ATP, B_1, folic acid, biotin, CoA, K, B_6). The bigger concentration of NAD and FMN than other coenzymes increase metabolism and produce fast saturation with H+ ions. The consequence is that these prokaryotic cells have high frequency of depolarization.

Prokaryotic cells with low frequency of depolarization produce the big complex metabolites if they are synthetic or big simple metabolites if they are decomposition because prokaryotic cells with low frequency of depolarization have enough time to build big complex and big simple molecules. The big complex molecules mean huge molecules of polysaccharides, peptidoglycan, and fatty acids. The big simple molecules mean the small presence of products of glycolysis and Krebs cycle and the big presence of monosaccharide and disaccharide molecules.

Prokaryotic cells with low frequency of depolarization have less concentrations of NAD and FMN coenzymes than do other coenzymes (ATP, B_1, folic acid, biotin, CoA, K, B_6). The less concentration of NAD and FMN than other coenzymes have increases metabolism slowly and produces slow saturation with H+ ions. The consequence is that these prokaryotic cells have low frequency of depolarization.

Prokaryotic cells with moderate frequency of depolarization produce moderately big complex metabolites if they are synthetic or moderately big simple metabolites if they are decomposition because they have enough time to do so. Moderately big complex molecules include polysaccharides, peptidoglycan, and fatty acids. The moderate big simple molecules mean the similar presence of products of glycolysis, Krebs cycle, and small monosaccharide and disaccharide molecules.

Prokaryotic cells with frequency of depolarization between the two extremes have concentrations of NAD and FMN coenzymes similar to other coenzymes (ATP, B_1, folic acid, biotin, CoA, K, B_6). Such similar concentrations increase metabolism and H+ saturation moderately and account for moderate frequency of depolarization.

Concentrations of metabolites outside cells depend on concentrations of coenzymes and metabolites attracted by polypeptides of protein spirals of prokaryotic cells or their power of metabolism. Concentration of metabolites are less than in prokaryotic cells if they have bigger speeds of metabolism and vice versa.

The different complex compounds of the standard concentration in the wells of the microtiter plate determine the specific enzymes and their quantities. The speed of metabolism under influence of the different complex compounds of the complex concentrations can be used for the presence of the different enzymes and their quantities.

Lactase	Galact-ase	Mann-ase	Saccha-rase	Lipase	Cellul-ase	Starch enzymes	PAH-ase	Urease	PET-ase	Collag-enase	DNA-ase

Different prokaryotic cells cannot be investigated by methods of practical taxonomy if they do not grow in natural media. This is the biggest challenge for practical taxonomy. They can be investigated only by analysis of their nucleic acids and genes. The results of the analysis of their nucleic acids and genes will be translated into polypeptides of protein spirals and data of the new taxonomy. Therefore, practical taxonomy will look for new technical devices and methods that will enable investigation of nucleic acids and genes of prokaryotic cells and translation of their data into data of the practical taxonomy.

Designing of different prokaryotic ecosystems will occur according to the characteristics of prokaryotic species of the practical taxonomy and the stored prokaryotic cells. The designing of different prokaryotic systems will be done also according to

- the chemical characteristics of the environment (prokaryotic cells must utilize different compounds from an environment in order to survive in it),
- the goal to decompose the specific compounds or increase specific from existing compounds, and
- the interchange of the different metabolites among prokaryotic cells of the designed prokaryotic ecosystem.

The design must ensure the stable prokaryotic ecosystem that will perform the goal for a long time.

Appendix XLII

This is an example of a proposal of the method for a treatment of malignant tumors. The proposed method relies on the hypotheses that normal and malignant cells have different frequency of depolarization, voltage, and current and that ultrasound can destroy malignant cells with the resonance phenomenon without killing normal cells. The destruction of malignant cells should rely on the resonance phenomenon between ultrasound and malignant cells because the resonance phenomenon exists for any object in nature. The literature regarding the different frequency of depolarization, voltage, and current of normal and malignant cells is very poor because technical devices are not used to determine those factors.

In "Bioelectrical Regulation of Cell Cycle and the Planarian Model System" (2015), Paul G. Barghouth, Manish Thiruvalluvan, and Nestor J. Oviedo[116] concluded,

Cell cycle regulation through the manipulation of endogenous membrane potentials offers tremendous opportunities to control cellular processes during tissue repair and cancer formation ... Although the characterization of such bioelectrical phenomena is well established, very little is known about how endogenous electric fields actually affect biological functions and the mechanism through which cells respond to their influence.

Literature about ultrasound used to treat diseases including malignancies is very rich. Data about the destruction of malignancies by ultrasound using the resonance phenomenon are not found in the contemporary literature.[117] Without these data, the resonance phenomenon using ultrasound for treating malignant tumors cannot be proven.

A proposal for the detection of different frequencies of depolarization, voltage, and currents of normal and malignant cells and delivery of ultrasound is next. The sliding electrodes in the antimicrobial tubes connect a malignant tumor, an oscilloscope, and a function generator in a specific way. The electrodes are inside of a tumor, and they do not attach each other.

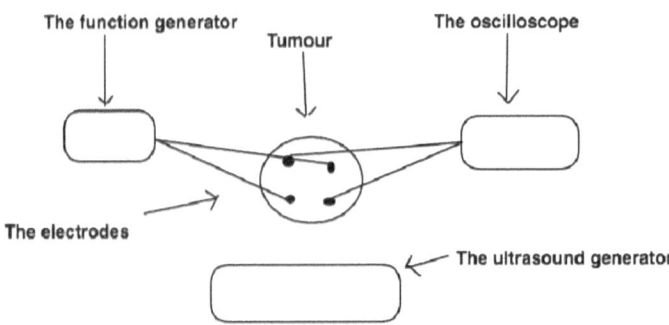

A function generator delivers weak electricity of constant voltage and current but at different frequencies to a tumor by sliding electrodes. An oscilloscope detects the electricity and voltage and different frequencies also by sliding electrodes because malignant cells are embedded in electrolytes that conduct electricity. If low-frequency electricity is the same as the bioelectricity of the tumor, the voltage and current detected by the oscilloscope must increase due to added voltage and current of the malignant cells' bioelectricity. This hypothesis can be proven or disproven only through experimentation. If an oscilloscope and a function generator can find frequencies of bioelectricity of malignant cells, we can try to destroy tumors that way and not damage normal cells.

References

1. *Jawetz, Melnick & Adelberg's Medical Microbiology* 24th edition, 23.
2. https://en.wikipedia.org/wiki/Enzyme.
3. "Viscous Dynamics of Lyme Disease and Syphylis Spirochetes Reveal Flagellar Torque and Drag," Michael Harman, Dhruv K. Vig, Justin D. Radolf, Charles W. Wolgemuth.
4. https://en.wikipedia.org/wiki/Cofactor_(biochemistry).
5. https://en.wikipedia.org/wiki/Citric_acid_cycle.
6. https://en.wikipedia.org/wiki/Fatty_acid_metabolism.
7. Lash, Timothy D. (2011). "Origin of aromatic character in porphyrinoid systems." *Journal of Porphyrins Phthalocyanines* 15: 1093–1115.
8. "Rules of engagement: centrosome–nuclear connections in a closed mitotic system," Meredith Leo, Diana Santino, Irina Tikhonenko, Valentin Magidson, Alexey Khodjakov, Michael P. Koonce, *Biology Open* 2012, 1: 111117; doi: 10.1242/bio.20122188.
9. Endoplasmic reticulum, Wikipedia, https://en.wikipedia.org/wiki/Endoplasmic_reticulum.
10. Golgi apparatus, Wikipedia, https://en.wikipedia.org/wiki/Golgi_apparatus.
11. *Richard Brooke Roberts, 1910–1980, A Biographical Memoir* by Roy J. Britten; ribosome, Wikipedia.
12. Proto-oncogenes, Wikipedia.
13. https://www.pnas.org/content/pnas/87/12/4576.full.pdf
    Towards a natural system of organisms: Proposal for the domains Archaea, Bacteria, and Eucarya, Carl R. Woese, Otto Kandler, and Mark L. Wheelis (1990).
14. "Virus taxonomy: classification and nomenclature of viruses," seventh report of the International Committee on Taxonomy

of Viruses. M. H. Regenmortel, M. H. V. van; C. M. Fauquet, et al. Book: *Virus taxonomy: classification and nomenclature of viruses. Seventh report of the International Committee on Taxonomy of Viruses* 2000, xii, 1162.

15. http://www.web-books.com/MoBio/Free/Ch1E2.htm.
16. *The Immortal Life of Henrietta Lacks* reviewed by Faroque A Khan, MB; MACP, Associate Editor, JIMA, *The Immortal Life of Henrietta Lacks* by Rebecca Skloot. Crown Publishers, 2010.
17. https://en.wikipedia.org/wiki/Malignant_transformation.
18. https://www.vocabulary.com/dictionary/hypothesis.
19. https://commons.wikimedia.org/wiki/File:Cancer_requires_multiple_mutations_from_NIHen.png.
20. https://en.wikipedia.org/wiki/Causes_of_cancer.
21. "DNA Damage, DNA repair and Cancer," Carol Bernstein, Anil Prasad, Valentine Nfonsan, and Harris Bernstein, 2013.
22. "Cellular signalling and Oncogenesis," Jacques Ghysdael Emiratus, Research Director, Institute Curie.
23. What is Cancer? https://www.cancer.gov/about-cancer/understanding/what-is-cancer https://en.wikipedia.org/wiki/Cancer.
24. Charlie Chaplin from a famous scene in *The Gold Rush*.
25. "Pathological changes in human malignant carcinoma treated with high-intensity focused ultrasound," F. Wu, W. Z. Chen, et al., https://www.ncbi.nlm.nih.gov/pubmed/11527596.
26. "High-intensity focused ultrasound in the treatment of solid tumours," J. E. Kennedy, https://www.ncbi.nlm.nih.gov/pubmed/?term=High-intensity+focused+ultrasound+in+the+treatment+of+solid+tumours+James+E.+Kennedy1.
27. "A review on the use of magnetic fields and ultrasound for non-invasive cancer treatment," S. Sengupta, [1,2] V. K. Balla.[1,2]
28. "Inhibition of Cancer Cell Growth by Exposure to a Specific Time-Varying Electromagnetic Field Involves T-Type Calcium Channels," Carly A. Buckner,[1,2] Alison L. Buckner,[1,2] Stan A. Koren,[3] Michael A. Persinger,[1,3] and Robert M. Lafrenie[1,2,4*]
29. https://en.wikipedia.org/wiki/Royal_Rife.
30. https://www.mcb.harvard.edu.

31. http://abbottg.wixsite.com/abbott-lab/bioelectricity.

32. *Happy Darwin Day. Will We Heed His Time-Tested Wisdom?* James Marshall Crotty, https://www.forbes.com/sites/jamesmarshallcrotty/2015/02/16/happy-darwin-day-will-we-heed-his-time-tested-wisdom/#2d95d992b06e.

33. James Watson, Francis Crick, Maurice Wilkinson, and Rosalind Franklin: Science History Institute, https://www.sciencehistory.org/historical-profile/james-watson-francis-crick-maurice-wilkins-and-rosalind-franklin.

34. https://en.wikipedia.org/wiki/Miroslav_Krleža.

35. "Electron microscopy of leptospires. I. Anatomical features of Leptospira pomona" (1965), A. E. Ritchie, National Animal Disease Laboratory, Ames, Iowa, and Herman C. Ellinghausen.

36. "Growth and cell division of Mycobacterium avium" (1981), N. Rastogi and H. L. David.

37. "Population genetics of a transformable bacterium: the influence of horizontal genetic exchange on the biology of Neisseria meningitides" (1993), M. C. Maiden.

38. "Hypothesis: chromosomes separation in Escherichia coli involves autocatalytic gene expression, transertion and membrane domain formation" (1995), V. H. Norris.

39. "Structure" (1996), M. R. J. Salton and K. S. Kim.

40. "Bipolar localization of the replication origin regions of chromosomes in vegetative and sporulation cells of Bacillus subtilis" (1997), C. D. Webb, A. Teleman A, S. Gordon, et al.

41. "Cell reproduction cycle of Mycoplasma" (1999), M Miyata Seto.

42. "Alternative to binary fission in bacteria" (2005), E. R. Angert.

43. "FtsZ and division of prokaryotic cells and organelles" (2005), W. Margolin.

44. "Duplication and Segregation of the actin (MreB) cytoskeleton during the prokaryotic cell cycle" (2007), Purva Vats and Lawrence Rothfield.

45. "Spatial complexity and control of a bacterial cycle" (2007), J. Colier and L. Shapiro.

46. "Ancient ESCRT and the evolution of binary fission (2009)," R. Y. Samson and S. D. Bell.

47. "Cell division in a minimal bacterium in the absence of ftsZ" (2010), M. Lluch-Senar, E. Querol, and J. Piñol.

48. "Bacterial chromosome organization and segregation" (2010), E. Toro and L. Shapiro.

49. "Factors affecting daughter cells' arrangement during the early bacterial divisions" (2010), P. T. Su, P. W. Yen, S. H. Wang, et al.

50. "Chromosome organization by a nucleoid-associated protein in live bacteria" (2011), W. Wang. G. W. Li, C. Chen, X. S. Xie, and X. Zhuang.

51. "Positive control of cell division: FtsZ is recruited by SsgB during sporulation of Streptomyces" (2011), J. Willemse, J. W. Borst, E. de Waal, et al.

52. "Molecular and structural basis of ESCRT-III recruitment to membranes during archaeal cell division" (2011), R. Y. Samson, T. Obita, B. Hodgson, et al.

53. "How did bacterial ancestors reproduce? Lessons from L-form cells and giant lipid vesicles: multiplications similarities between lipid vesicles and L-form bacteria" (2012), Y. Briers, P. Walde, M. Schuppler, and M. J. Loessner.

54. "The bacterial Min system" (2013), V. W. Rowlett, W. Margolin.

55. "Chromosome architecture is a key element of bacterial cellular organization" (2013), J. Ptacin, L. Shapiro.

56. "How myxobacteria glide" (2002), C. Wolgemuth, E. Hoiczyk, D. Kaiser, and G. Oster.

57. "The surprisingly diverse ways that prokaryotes move" (2008), K. F. Jarrel, M. J. McBride.

58. "Physics of Intracellular Organization in Bacteria" (2015), N. S. Wingreen and K. C. Huang.

59. https://www.youtube.com/watch?v=JA5j8QzAz0w.

60. https://www.youtube.com/watch?v=0VgNBl2JomQ. https://www.youtube.com/watch?v=0S0CWAmmFzY.

61. https://www.ncbi.nlm.nih.gov/pubmed/7968920

62. "Do prokaryotes contain microtubules?" D. Bermudes, G. Hinkle, and L Margulis.

63. The structure of the extracted bacterial flagellar motor was described in a seminal paper by Dennis Thomas, David Morgan, and David DeRosier (1999), "Rotational Symmetry of the C-ring and a Mechanism for the Flagellar Rotary Motor" (1999), Proc. Natl. Acad. Sci. USA *96*, 10134-10139).

64. *Structure and Function of Bacterial Cells*, Keneth Todar.

65. *Functional Anatomy of Prokaryotic and Eukaryotic Cells*, Bradley W. Christian, McLennon Community College.

66. "The stack: a new bacterial structure analyzed in the Antarctic bacterium Pseudomonas deceptionensis M1(T) by transmission electron microscopy and tomography," L. Delgado, O. Carrión, G. Martínez, C. López-Iglesias, and E. Mercadé, PLoS One (2013) Sep 9;8(9):e73297. doi:10.1371/journal.pone.0073297. eCollection 2013.

67. "Basic Structures of Prokaryotic Cells," Boundless Biology, January 2016, retrieved May 5.

68. *Principle of Biochemistry/ Cell and its Biochemistry*, https://en.wikibooks.org/wiki/Principles_of_Biochemistry/Cell_and_its_Biochemistry.

69. *The New Bacterial Cell Biology: Review Moving Part and Subcellular Architecture*, Zemer Gitai, Department of Developmental Biology Beckman Center School of Medicine, Stanford University; Department of Molecular Biology, Princeton University.

70. https://en.wikipedia.org/wiki/Redox.

71. https://en.wikipedia.org/wiki/Flavin_adenine_dinucleotide.

72. *The role of iron-porhyrin compounds in biological oxidations*, E. S. Guzman Barron, Cold Spring Harbour Symposia on Quantitative Biology.

73. "Transposons, or jumping genes: Not junk DNA?" (2008), L. Pray, *Nature Education* 1(1):32.

74. https://en.wikipedia.org/wiki/Messenger_RNA.

75. http://www.biologyexams4u.com/2013/04/steps-involved-in-dna-replication-in.html#.VShJ_16Gn1o.

76. https://en.wikipedia.org/wiki/Cytoskeleton.

77. *Microtubules in Bacteria: Ancient Tubulin Build a Five-Protofilament Homolog of the Eukaryotic Cytoskeleton*, Marin Pilhofer, Mark S. Ladinsky, Aladair W. McDowall, Giullio Petroni, and Grant. J. Jensen.

78. https://en.wikipedia.org/wiki/Mitochondrion.

79. https://embryo.asu.edu/pages/dna-and-x-and-y-chromosomes.

80. https://en.wikipedia.org/wiki/Chromosome.

81. https://en.wikipedia.org/wiki/DNA.

82. https://en.wikipedia.org/wiki/Cell_nucleus.

83. https://en.wikipedia.org/wiki/Nucleolus.

84. https://en.wikipedia.org/wiki/Centrosome.

85. https://en.wikipedia.org/wiki/Ribosome.

86. https://www.youtube.com/watch?v=Ikq9AcBcohA.

87. https://www.youtube.com/watch?v=5bLEDd-PSTQ.

88. https://www.youtube.com/watch?v=2BwWavExcFI.

89. https://www.britannica.com/science/mitosis.

90. https://commons.wikimedia.org/wiki/File:Nucleosome_structure-2.png.

91. https://en.wikipedia.org/wiki/Meiosis.

92. https://www.britannica.com/science/meiosis-cytology.

93. http://www.phschool.com/science/biology_place/labbench/lab3/concepts2.html.

94. http://www.macroevolution.net/prophase-i.html.

95. https://en.wikipedia.org/wiki/Regulation_of_gene_expression.

96. https://www.youtube.com/watch?v=3S3ZOmleAj0.

97. https://en.wikipedia.org/wiki/Bacteriophage.

98. https://en.wikipedia.org/wiki/Viral_replication.

99. "Hepatitis B virus replication," J. Beck and M. Nassal. "Hepadnavirus Genome Replication and Persistence," J. Hu and C. Seeger.

100. "Microbes Drives Evolution of Animals and Plants: the Hologenome Concept," Eugene Rosenberg, Ilana Zilberg-Rosenberg.

101. https://en.wikipedia.org/wiki/Mutation.

102. https://uncsom.wordpress.com.

103. "In search of a bacterial species definition" Edgardo Moreno (Programa de Investigación en Enfermedades Tropicales (PIET), Escuela de Medicina Veterinaria, Universidad Nacional, Heredia, Costa Rica.

104. "What are bacterial species?" F. M. Cohan, https://www.ncbi.nlm.nih.gov/pubmed/12142474.

105. "Phylogenomic Analyses and Reclassification of Species within the Genus Tsukamurella: Insights to Species Definition in the Post-genomic Era," J. L. Teng, Y. Tang, Y. Huang, et al., https://www.ncbi.nlm.nih.gov/pmc/articles/PMC4955295/.

106. https://kyocp.wordpress.com/2012/06/14/types-of-bacteria-used-in-wastewater-treatment/.

107. "Drivers and applications of integrated clean-up technologies for surfactant-enhanced remediation of environments contaminated with polycyclic aromatic hydrocarbons (PAHs)," X. Liang, C. Guo, C. Liao, et al.

108. "Beneficial Microorganisms for Corals (BMC): Proposed Mechanisms for Coral Health and Resilience," R. S. Peixoto, P. M. Rosado, D. C. Leite, et al.

109. "Degradation of textile dyes by cyanobacteria," P. M. Dellamatrice, M. E. Silva-Stenico, L. A. Moraes, et al., https://www.ncbi.nlm.nih.gov/pubmed/28341397.

110. "Metagenomic analysis exploring taxonomic and functional diversity of soil microbial communities in Chilean vineyards and surrounding native forests," L. E. Castañeda and O. Barbosa, https://www.ncbi.nlm.nih.gov/pubmed/28382231.

111. "Lantibiotics produced by Actinobacteria and their potential applications (a review)," K. M. Gomes, R. S. Duarte, and Bastos de Freire, https://www.ncbi.nlm.nih.gov/pubmed/28270262.

112. http://www.yourarticlelibrary.com/bacteria/importance-of-bacteria-to-agriculture-and-industries-1049-words/6789/.

113. https://www.sigmaaldrich.com/catalog/product/sial/b2551?lang=en&region=US.

114. https://www.epa.gov/sites/production/files/2014-03/documents/pahs_factsheet_cdc_2013.pdf; "Biodegradation aspects of Polycyclic Aromatic Hydrocarbons (PAHs): A

review," A. K. Haritash and C. P. Kaushik, Departments of Environmental Science & Engineering, Guru Jambheshwar University of Science & Technology, Hisar, Haryana, India, https://marine.rutgers.edu/dmcs/ms606/2010_fall/Haritash2009.pdf.

115. https://www.azocleantech.com/article.aspx?ArticleID=254, https://aem.asm.org/content/aem/early/2018/01/29/AEM.02773-17.full.pdf, https://www.ncbi.nlm.nih.gov/pubmed/29427431. "New Insight into the Function and Global Distribution of Polyethylene Terephthalate (PET)—Degrading Bacteria and Enzymes in Marine and Terrestrial Metagenomes," D. Danso, C. Schmeisser, J. Chow, et al.

116. "Bioelectrical Regulation of Cell Cycle and the Planarian Model System," Paul G. Barghouth, Manish Thiruvalluvan, and Nestor J. Oviedo, https://www.ncbi.nlm.nih.gov/pmc/articles/PMC4561208/.

117. "Altering bioelectricity on inhibition of human breast cancer cells," Seher Berzingi, Mackenzie Newman, and Han-Gang Yu, https://www.ncbi.nlm.nih.gov/pmc/articles/PMC5034549/.

118. https://simple.wikipedia.org/wiki/Enzyme

119. https://en.wikipedia.org/wiki/Polyethylene_terephthalate

120. https://www.ncbi.nlm.nih.gov/pmc/articles/PMC2009334/ Lymphoreticular infiltration in human tumours: prognostic and biological implications: a review.
J. C. Underwood

121. https://www.truerife.com/products-2.html

www.ingramcontent.com/pod-product-compliance
Lightning Source LLC
Chambersburg PA
CBHW020722180526
45163CB00001B/72